Communications
Formulas & Algorithms

Other McGraw-Hill Communications Books of Interest

AZEVEDO · *ISPF*
BERSON · *APPC: A Guide to LU6.2*
CHORAFAS · *The Complete Lan Reference*
COOPER · *Computer and Communications Security*
DAYTON (RANADE, ED.) · *Integrating Digital Services*
FOLTS · *The McGraw-Hill Compilation of Data Communications Standards*
FORTIER · *Handbook of Lan Technology*
FTHENAKIS · *Manual of Satellite Communications*
HA · *Digital Satellite Communications*
HELD, SARCH · *Data Transmission*
INGLIS · *Electronic Communications Handbook*
KESSLER · *ISDN*
KNIGHTSON · *Standards for Open Systems Connection*
LEE · *Mobile Cellular Telecommunications Systems*
LEE · *Mobile Communications Engineering*
LUTZ · *Large Systems Networking with VTAM*
NEMZOW · *Keeping the Link*
OWEN · *Digital Transmission Systems*
RANADE · *Advanced SNA Networking*
RANADE · *Introduction to SNA Networking*
RANADE · *VSAM: Performance, Design, and Fine Tuning*
RANADE, RANADE · *VSAM: Concepts, Programming, and Design*
RHEE · *Error Correction Coding Theory*
ROHDE · *Communications Receivers*
SABIN · *Single-Sideband Systems and Circuits*
SARCH · *Integrating Voice and Data*
SARCH, ABBATIELLO · *Telecommunications & Data Communications Fact Book*
SCHLAR · *Inside X.25*
TUGAL · *Data Transmission*
UNGARO · *Networking Software*

Subscription information to BYTE Magazine:
Call 1-800-257-9402 or write Circulation Dept.,
One Phoenix Mill Lane, Peterborough, NH 03458.

Communications Formulas & Algorithms

For Systems Analysis and Design

C. Britton Rorabaugh

McGraw-Hill, Inc.
New York   St. Louis   San Francisco   Auckland   Bogotá
Caracas   Hamburg   Lisbon   London   Madrid
Mexico   Milan   Montreal   New Delhi   Paris
San Juan   São Paulo   Singapore
Sydney   Tokyo   Toronto

Library of Congress Cataloging-in-Publication Data

Rorabaugh, C. Britton.
  Communications formulas & algorithms : for systems analysis and
design / C. Britton Rorabaugh.
      p.   cm.
  Includes bibliographical references and index.
  ISBN 0-07-053644-9
  1. Telecommunication systems—Design and construction—Handbooks,
manuals, etc. 2. Telecommunication—Mathematics—Handbooks,
manuals, etc. I. Title.   II. Title: Communications formulas and
algorithms.
TK5102.5.R56   1990
621.382′011—dc20
                                                            90-36552
                                                                 CIP

Copyright © 1990 by McGraw-Hill, Inc. All rights reserved. Printed in the United States of America. Except as permitted under the United States Copyright Act of 1976, no part of this publication may be reproduced or distributed in any form or by any means, or stored in a data base or retrieval system, without the prior written permission of the publisher.

1 2 3 4 5 6 7 8 9 0   DOC/DOC   9 5 4 3 2 1 0

ISBN 0-07-053644-9

The sponsoring editor for this book was Daniel A. Gonneau, the editing supervisor was Marci Nugent, the designer was Naomi Auerbach, and the production supervisor was Dianne Walber. It was set in Century Schoolbook by Datapage.

Printed and bound by R. R. Donnelley & Sons Company.

Information contained in this work has been obtained by McGraw-Hill, Inc., from sources believed to be reliable. However, neither McGraw-Hill nor its authors guarantees the accuracy or completeness of any information published herein and neither McGraw-Hill nor its authors shall be responsible for any errors, omissions or damages arising out of use of this information. This work is published with the understanding that McGraw-Hill and its authors are supplying information but are not attempting to render engineering or other professional services. If such services are required, the assistance of an appropriate professional should be sought.

TAB BOOKS offers software for
sale. For information and a catalog,
please contact TAB Software Department,
Blue Ridge Summit, PA 17294-0850.

To Joyce, Geoff, and Amber

Contents

Preface xv

Chapter 1. Introduction to Communications   1

1.1 Historical Roots   1
1.2 References   1

Chapter 2. Mathematics Review   3

2.1 Exponentials and Logarithms   3
    Exponentials   3
    Logarithms   4
    Decibels   4
2.2 Trigonometry   6
    Phase Shifting of Sinusoids   6
    Trigonometric Identities   7
    Euler's Identities   8
    Series and Product Expansions   8
    Orthonormality of Sine and Cosine   9
2.3 Complex Numbers   11
    Operations on Complex Numbers in Rectangular Form   12
    Polar Form of Complex Numbers   12
    Operations on Complex Numbers in Polar Form   13
    Logarithms of Complex Numbers   14
2.4 Derivatives   14
2.5 Integration   15
2.6 Dirac Delta Function   17
    Distributions   19
    Properties of the Delta Distribution   19
2.7 Bessel Functions   20
    Bessel Functions of the First Kind   20
    Bessel Function Identities   20
    Modified Bessel Functions of the First Kind   22
    Identities for Modified Bessel Functions   22
    Evaluation of Bessel Functions   23

2.8 Special Functions — 23
Gamma Function — 23
Confluent Hypergeometric Function — 24
Error Function — 24
Q Function — 25
Marcum Q Function — 27
2.9 Combinations and Permutations — 27
2.10 References — 28

Chapter 3. Probability and Random Variables — 29

3.1 Randomness and Probability — 29
Joint and Conditional Probabilities — 30
Independent Events — 31
3.2 Bernoulli Trials — 31
3.3 Random Variables — 32
Cumulative Distribution Functions — 33
Properties of Distribution Functions — 33
Probability Density Function — 33
Types of Random Variables — 34
3.4 Moments of a Random Variable — 34
Mean — 34
Mean of a Function of a Random Variable — 35
Moments — 36
Central Moments — 36
Properties of Variance — 36
Absolute and General Moments — 37
3.5 Characteristic Functions — 37
3.6 Relationships between Random Variables — 38
Statistical Independence — 39
Conditional Probability Densities — 39
Joint Moments — 40
3.7 Correlation and Covariance — 40
3.8 Probability Densities for Functions of a Random Variable — 41
3.9 Probability Densities for Functions of Two Random Variables — 42
Probability Density Function for a Linear Combination of Two Random Variables — 42
Probability Distribution for a Product of Discrete Random Variables — 43
3.10 References — 43

Chapter 4. Probability Distributions in Communications — 45

4.1 Gaussian Random Variables — 45
4.2 Uniform Random Variables — 48
4.3 Exponential Random Variables — 49
4.4 Rayleigh Random Variables — 49
4.5 Rice Random Variables — 52
4.6 Chi-Squared Random Variables — 54
4.7 Poisson Distribution — 55
4.8 Gamma and Erlang Distributions — 55
4.9 References — 55

Chapter 5. Random Processes — 57

- 5.1 Random Processes — 57
- 5.2 Stationarity — 59
- 5.3 Autocorrelation and Autocovariance — 60
  - Properties of Autocorrelation Functions — 61
  - Autocovariance — 61
  - Uncorrelated Random Processes — 61
- 5.4 Power Spectral Density of Random Processes — 62
- 5.5 Linear Filtering of Random Processes — 62
- 5.6 Gaussian Random Processes — 63
  - Conditional pdf of Two Sample Functions from a Gaussian Random Process — 63
  - Joint pdf of Two Sample Functions from a Gaussian Random Process — 64
- 5.7 Markov Processes — 64
  - Markov Chains — 64
  - Classification of Markov Chains — 65
  - State Diagrams of Markov Chains — 66
  - Multistep Transition Probabilities — 68
  - Absolute State Probabilities — 69
  - First Passage Time — 69
  - Classification of States — 70
  - Reducibility of Markov Chains — 70
  - Final State Probabilities — 71
- 5.8 References — 72

Chapter 6. Signals and Spectra — 73

- 6.1 Mathematical Modeling of Signals — 73
- 6.2 Energy Signals versus Power Signals — 76
- 6.3 Fourier Series — 77
  - Trigonometric Forms — 77
  - Exponential Form — 78
  - Conditions of Applicability — 79
  - Properties of the Fourier Series — 80
  - Fourier Series of a Square Wave — 82
  - Parseval's Theorem — 84
- 6.4 Fourier Transform — 84
  - Fourier Transforms of Periodic Signals — 86
  - Common Fourier Transform Pairs — 87
- 6.5 Spectral Density — 87
  - Energy Spectral Density — 87
  - Power Spectral Density of a Periodic Signal — 87
- 6.6 Autocorrelation — 90
  - Autocorrelation of Energy Signals — 90
  - Autocorrelation of Power Signals — 91
- 6.7 Cross-Correlations — 91
  - Cross-Correlation of Energy Signals — 91
  - Cross-Correlation of Power Signals — 92
- 6.8 Bandpass Signals — 92
  - Complex Envelope — 93
  - Preenvelope — 94
- 6.9 References — 94

Chapter 7. System Theory 97

- 7.1 Systems — 97
  - Linearity — 98
  - Time Invariance — 100
  - Causality — 100
- 7.2 Characterization of Linear Systems — 101
  - Impulse Response — 101
  - Step Response — 102
- 7.3 Laplace Transform — 103
- 7.4 Properties of the Laplace Transform — 105
  - Time Shift Right — 105
  - Time Shift Left — 106
- 7.5 Transfer Functions — 107
- 7.6 Heaviside Expansion — 110
  - General Case — 110
  - Simple Pole Case — 110
- 7.7 Poles and Zeros — 111
- 7.8 Magnitude, Phase, and Delay Responses — 113
  - Phase Delay — 113
  - Group Delay — 114
- 7.9 Bandpass Systems — 115
- 7.10 References — 116

Chapter 8. Noise 117

- 8.1 White Noise — 117
- 8.2 Noise Equivalent Bandwidth — 118
- 8.3 Thermal Noise — 119
  - Equivalent Noise Temperature — 120
  - Noise Figure — 121
- 8.4 Quadrature Form Representation of Bandpass Noise — 121
  - General Case — 121
  - Gaussian Case — 123
- 8.5 Superposition of Noise Powers — 123
- 8.6 Amplitude of a Sine Wave with Random Phase — 123
- 8.7 Amplitude of a Sine Wave Plus Gaussian Noise — 124
- 8.8 Envelope of a Sine Wave Plus Bandpass Gaussian Noise — 126
- 8.9 Squared Envelope of Bandpass Gaussian Noise — 128
- 8.10 Phase of a Sine Wave Plus Bandpass Gaussian Noise — 128
- 8.11 Postdetection Integration of Bandpass Gaussian Noise — 130
- 8.12 References — 130

Chapter 9. Communication Channels 133

- 9.1 Radio Spectrum — 133
- 9.2 Propagation — 134
- 9.3 Ground Wave Propagation — 135
  - Surface Wave Propagation — 135
  - Space Wave — 136
  - Line-of-Sight — 136

9.4	Additive Gaussian Noise Channel	137
9.5	Band-Limited Channel	138
9.6	References	139

Chapter 10. Detection Theory — 141

10.1	Binary Decision Problem	142
	General Case	142
	Binary Decisions Using a Single Division Variable	143
10.2	Optimal Decision Criteria	146
	Maximum Likelihood Decision Criterion	146
	Bayes Decision Criterion	147
	Ideal Observer Decision Criterion	147
	Neyman-Pearson Decision Criterion	147
10.3	Optimum Coherent Detection of Binary Signals in the AWGN Channel	148
	Antipodal Signals	152
	Orthogonal Signals	153
10.4	Optimum Coherent Detection of M-ary Signals in the AWGN Channel	153
	Orthogonal Signals	154
	Biorthogonal Signals	155
	Equicorrelated Signals	157
10.5	Optimum Noncoherent Detection in an AWGN Channel	159
	Binary Signals	159
	M-ary Orthogonal Signals	161
10.6	References	161

Chapter 11. Signal Processing and Simulation Issues — 163

11.1	Digital Processing in Communication Systems	163
	Digitization	163
	Ideal Sampling	165
	Sampling Rate Selection	166
	Instantaneous Sampling	167
	Natural Sampling	169
11.2	Discrete Fourier Transform	171
11.3	Simulation of White Gaussian Noise	172
	Method A	172
	Method B	173
11.4	Variance for Simulation of White Noise	174
11.5	Simulation of Noise in the Frequency Domain	175
11.6	References	177

Chapter 12. Continuous Modulation — 179

12.1	Amplitude Modulation	179
	Spectrum of AM Signals	180
	Power in an AM Signal	181
	Square Wave Modulation	182
	Sinusoidal Modulation	183
12.2	Modulators for AM	183
	Square-Law Modulator	183
	Switching Modulator	185

12.3	Measures of Demodulator Performance	187
12.4	Envelope Demodulation	188
	Demodulation Gain	189
	Figure of Merit	191
12.5	Square-Law Detector	193
12.6	Coherent Demodulation of AM Signals	194
12.7	Double-Sideband Suppressed-Carrier (DSBSC) Modulation	195
	Spectrum of DSBSC Signals	196
	Generation of DSBSC Using a Balanced Modulator	196
	Demodulation of DSBSC Signals	197
12.8	Single Sideband	198
	SSB Modulation via Frequency Discrimination	200
	Generation of SSB Signals Using a Hartley Modulator	200
	Demodulation of SSB Signals	202
12.9	Angle Modulation	202
	Single-Tone Angle Modulation	203
	Double-tone Angle Modulation	204
	Multitone Angle Modulation	204
	Narrowband FM	204
12.10	References	205

Chapter 13. On-Off Keying 207

13.1	Noncoherent Detection of OOK	207
	Performance	208
	Optimum Threshold	213
13.2	Coherent Detection of OOK	214
	Performance	214
	Optimum Threshold	218
13.3	References	218

Chapter 14. Frequency-Shift Keying 219

14.1	Binary Frequency-Shift Keying	219
14.2	Continuous-Phase Frequency-Shift Keying (CPFSK)	221
14.3	Coherent Detection of FSK Signals	226
	Orthogonal Signals	226
	Nonorthogonal Signals	227
14.4	Noncoherent Detection of FSK	228
	Bandpass Filter Approach	228
	Matched-Filter Approach	230
	Nonorthogonal Signals	230
14.5	M-ary FSK	231
14.6	Detection of MFSK Signals	234
	Coherent Detection	234
	Noncoherent Detection	237
14.7	Binary FSK Performance in Rayleigh Fading	239
	Coherent Detection	239
	Noncoherent Detection	239
14.8	Diversity Performance of Binary Orthogonal FSK	240
14.9	M-ary FSK Performance in Rayleigh Fading	240
14.10	References	242

Chapter 15. Phase-Shift Keying   243

   15.1   Binary Phase-Shift Keying   243
   15.2   Phase Reversal Keying   246
   15.3   Geometric Representation of PSK Signals   247
   15.4   QPSK   248
   15.5 M-ary Phase-Shift Keying   248
   15.6   References   250

Index   251

Preface

Much of the electrical engineering literature—both journal articles and books—is written by academics for academics. Most of the useful information is buried within pages of rigorous derivation, formal proofs, and meticulous expressions of concern over pathological cases which never surface in practical work. These shortcomings are further compounded by the different variations of notation and terminology often used by different authors to discuss similar concepts. Consequently, a student or working engineer has a difficult time finding, understanding, and employing information on the most powerful theories and techniques which could be applied to his or her specific problem. This situation is particularly bad in the area of communication theory.

I have attempted to improve upon the existing literature by writing this book in a format and style which make it easier for a student or working engineer to locate, understand, and employ the appropriate techniques, properties, and theories. It is assumed that the reader is a student, expert hobbyist, or practicing professional in electrical or systems engineering. However, the reader is only human, and as such, prone to forget the details of most things not used on a regular basis. Therefore, all important details and *pertinent* mathematical background are included. Most of the mathematical presentations are distillations and clarifications of material available elsewhere only in a much less "user-friendly" form. Some of this material comes from statistical and mathematical literature not often available to, or consulted by, practitioners outside of the academic community. In cases where different and conflicting definitions or notational schemes are prevalent, an attempt has been made to point out and explain the various possibilities which a reader may encounter when consulting other sources. This book is not intended to be a textbook or a first introduction to communication theory, but I hope it proves to be an invaluable "one-stop" reference that readers reach for first.

Britt Rorabaugh

Chapter 1

Introduction to Communications

A radiotelegraph system, wireline data communication system, and fiber-optic local area network have widely varying characteristics; but each of these systems can be partitioned into message sources, transmitters, channels, and receivers. Communication theory is concerned with the *conceptual* analysis and design of each of these components in such a way that together they form a workable system.

Much of communication theory is mathematically intensive, and a great deal of the space in most communications books is devoted to mathematical derivations. While certainly interesting and tutorially useful, such material is as helpful to most communications practitioners as a treatise on hammer design would be to a carpenter. The equations and algorithms in this book are tools for communications engineers, just as hammers and saws are tools for carpenters. The aim of this book is to provide the tools and information concerning their use without unnecessarily belaboring their origins. There is one exception—biographical information about some pioneers in the field is included just because it's fun stuff to know and because it conveys a sense of just how far back the roots of modern communications extend.

1.1 Historical Roots

Table 1.1 list a number of early contributors whose names are attached to theorems and such used in this book.

1.2 References

1. I. Asimov: *Asimov's Biographical Encyclopedia of Science and Technology*, Doubleday, Garden City, N.Y., 1964.
2. C. B. Boyer: *A History of Mathematics*, Wiley, New York, 1968.

TABLE 1.1 Landmark Contributors to Communication Theory

Lifetime	Contributor	Occupation	Contribution
1550–1617	John Napier	Scottish mathematician	Napierian logarithms
1654–1705	Jacques (or Jakob) Bernoulli	Swiss mathematician	Bernoulli numbers
1667–1754	Abraham Demoivre	French mathematician	Demoivre's theorem
1700–1782	Daniel Bernoulli	Swiss mathematician	Bernoulli trials
1707–1783	Leonhard Euler	Swiss mathematician	Euler's identities
			Euler numbers
1749–1827	Pierre Simon Laplace	French mathematician	Laplace transform
1768–1822	Jean Robert Argand	French mathematician	Argand diagram
1768–1830	Jean Baptise Joseph Fourier	French mathematician	Fourier series
			Fourier transform
1777–1855	Johann Karl Friedrich Gauss	German mathematician	Gaussian distribution
1784–1846	Friedrich Wilhelm Bessel	German astronomer	Bessel functions
1789–1857	Augustin Louis Cauchy	French mathematician	Cauchy-Riemann equations
			Cauchy mean value theorem
1805–1859	Peter Gustav Lejeune Dirichlet	Mathematician	Dirichlet conditions
1810–1893	Ernst Eduard Kummer	German mathematician	Kummer's transformations
1823–1891	Leopold Kronecker	German mathematician	Kronecker delta
1826–1866	Georg Friedrich Bernhard Riemann	German mathematician	Riemann sum
			Cauchy-Riemann equations
1842–1919	Lord Rayleigh (John William Strut)	English physicist	Rayleigh distribution
			Rayleigh's energy theorem
1850–1925	Oliver Heaviside	English mathematician	Heaviside expansion
1856–1904	Andrei Andreyevich Markov	Russian mathematician	Markov processes
			Markov chains
1862–1943	David Hilbert	German mathematician	Hilbert transform
1871–1956	Emile Borel	French mathematician	Borel field
1894–1964	Norbert Wiener	American mathematician	Wiener-Khintchine theorem
1902–1984	Paul Adrien Murice Dirac	English physicist	Dirac delta function
1916–	Claude Elwood Shannon	American mathematician	Shannon's sampling theorem

Chapter 2

Mathematics Review

2.1 Exponentials and Logarithms

Exponentials

There is an irrational number, usually denoted as e, that is of great importance in virtually all fields of science and engineering. This number is defined by

$$e \triangleq \lim_{x \to +\infty} \left(1 + \frac{1}{x}\right)^x \approx 2.71828 \cdots \qquad (2.1)$$

Unfortunately, this constant remains unnamed and writers are forced to settle for calling it "the number e" or perhaps the "base of natural logarithms." The letter e was first used to denote the irrational in (2.1) by Leonhard Euler (1707–1783), so it would seem reasonable to refer to the number under discussion as "Euler's constant." Such is not the case, however, as the term *Euler's constant* is attached to the constant γ defined by

$$\gamma = \lim_{N \to \infty} \left(\sum_{n=1}^{N} \frac{1}{n} - \log_e N\right) \approx 0.577215664 \cdots \qquad (2.2)$$

The number e is most often encountered in situations where it is raised to some real or complex power. The notation $\exp(x)$ is often used in place of e^x since the former can be written more clearly and typeset more easily than the latter—especially in cases where the exponent is a complicated expression rather than just a single variable. The value for e raised to a complex power z can be expanded in an infinite series as:

$$\exp(z) = \sum_{n=0}^{\infty} \frac{z^n}{n!} \tag{2.3}$$

The series in (2.3) converges for all complex z having finite magnitude.

Logarithms

The *common logarithm*, or *base 10 logarithm*, of a number x is equal to the power to which 10 must be raised in order to equal x:

$$y = \log_{10} x \Leftrightarrow x = 10^y \tag{2.4}$$

The *natural logarithm*, or *base e logarithm*, of a number x is equal to the power to which e must be raised in order to equal x:

$$y = \log_e x \Leftrightarrow x = \exp(y) = e^y \tag{2.5}$$

Natural logarithms are also called *napierian* logarithms in honor of John Napier (1550–1617), a Scottish amateur mathematician who in 1614 published the first account of logarithms in *Mirifici logarithmorum canonis descripto (A Description of the Marvelous Rule of Logarithms)* [Ref. 2]. The concept of logarithms can be extended to any positive base b, with the base b logarithm of a number x equaling the power to which the base must be raised in order to equal x:

$$y = \log_b x \Leftrightarrow x = b^y \tag{2.6}$$

The notation log without a base explicitly indicated usually denotes a common logarithm, although sometimes this notation is used to denote natural logarithms (especially in some of the older literature). More often, the notation ln is used to denote a natural logarithm.

Logarithms exhibit a number of properties which are listed in Table 2.1. Entry 1 is sometimes offered as the definition of natural logarithms. The multiplication property in entry 3 is the theoretical basis upon which the design of the sliderule is based.

Decibels

Consider a system which has an output power of P_{out} and an output voltage of V_{out} given an input power of P_{in} and an input voltage of V_{in}. The gain, in decibels (dB), of the system is given by

$$G_{(dB)} = 10 \log_{10}\left(\frac{P_{out}}{P_{in}}\right) = 10 \log_{10}\left(\frac{V_{out}^2/Z_{out}}{V_{in}^2/Z_{in}}\right) \tag{2.7}$$

TABLE 2.1 Properties of Logarithms

1. $\ln x = \int_1^x \frac{1}{y} dy \quad x > 0$

2. $\frac{d}{dx}[\ln x] = \frac{1}{x} \quad x > 0$

3. $\log_b(xy) = \log_b x + \log_b y$

4. $\log_b\left(\frac{1}{x}\right) = -\log_b x$

5. $\log_b(y^x) = x \log_b y$

6. $\log_c x = (\log_b x)(\log_c b) = \dfrac{\log_b x}{\log_b c}$

7. $\ln(1+z) = \sum_{n=1}^{\infty} (-1)^{n-1} \dfrac{z^n}{n} \quad |z| < 1$

If the input and output impedances are equal, (2.7) reduces to

$$G_{(\text{dB})} = 10 \log_{10}\left(\frac{V_{\text{out}}^2}{V_{\text{in}}^2}\right) = 20 \log_{10}\left(\frac{V_{\text{out}}}{V_{\text{in}}}\right) \tag{2.8}$$

Example An amplifier has a gain of 17.0 dB. For a 3-mW input, what will the output power be?

solution Substituting the given data into (2.7) yields

$$17.0 \text{ dB} = 10 \log_{10}\left(\frac{P_{\text{out}}}{3 \times 10^{-3}}\right)$$

Solving for P_{out} then produces

$$P_{\text{out}} = (3 \times 10^{-3}) 10^{(17/10)} = 1.5 \times 10^{-1} = 150 \text{ mW}$$

Example What is the range in decibels of the values which can be represented by an 8-bit unsigned integer?

solution The smallest value is 1 and the largest value is $2^8 - 1 = 255$. Thus

$$20 \log_{10}\left(\frac{255}{1}\right) = 48.13 \text{ dB}$$

The abbreviation dBm is used to designate power levels relative to 1 mV. For example:

$$30 \text{ dBm} = 10 \log_{10}\left(\frac{P}{10^{-3}}\right)$$

$$P = (10^{-3})(10^3) = 10^0 = 1.0 \text{ W}$$

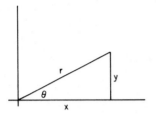

Figure 2.1 An angle in the cartesian plane.

2.2 Trigonometry

For x, y, r, and θ as shown in Fig. 2.1, the six trigonometric functions of the angle θ are defined as:

$$\text{sine:} \quad \sin\theta = \frac{y}{r} \tag{2.9}$$

$$\text{cosine:} \quad \cos\theta = \frac{x}{r} \tag{2.10}$$

$$\text{tangent:} \quad \tan\theta = \frac{y}{x} \tag{2.11}$$

$$\text{cosecant:} \quad \csc\theta = \frac{r}{y} \tag{2.12}$$

$$\text{secant:} \quad \sec\theta = \frac{r}{x} \tag{2.13}$$

$$\text{cotangent:} \quad \cot\theta = \frac{x}{y} \tag{2.14}$$

Phase shifting of sinusoids

A number of useful equivalences can be obtained by adding particular phase angles to the arguments of sine and cosine functions:

$$\cos(\omega t) = \sin\left(\omega t + \frac{\pi}{2}\right) \tag{2.15}$$

$$\cos(\omega t) = \cos(\omega t + 2n\pi) \quad n = \text{any integer} \tag{2.16}$$

$$\sin(\omega t) = \sin(\omega t + 2n\pi) \quad n = \text{any integer} \tag{2.17}$$

$$\sin(\omega t) = \cos\left(\omega t - \frac{\pi}{2}\right) \tag{2.18}$$

$$\cos(\omega t) = -\cos[\omega t + (2n+1)\pi] \qquad n = \text{any integer} \qquad (2.19)$$

$$\sin(\omega t) = -\sin[\omega t + (2n+1)\pi] \qquad n = \text{any integer} \qquad (2.20)$$

Trigonometric Identities

The following trigonometric identities will often prove useful in the theoretical analysis of communication systems.

$$\tan x = \frac{\sin x}{\cos x} \qquad (2.21)$$

$$\sin(-x) = -\sin x \qquad (2.22)$$

$$\cos(-x) = \cos x \qquad (2.23)$$

$$\tan(-x) = -\tan x \qquad (2.24)$$

$$\cos^2 x + \sin^2 x = 1 \qquad (2.25)$$

$$\cos^2 x = \frac{1}{2}[1 + \cos(2x)] \qquad (2.26)$$

$$\sin(x \pm y) = (\sin x)(\cos y) \pm (\cos y)(\sin y) \qquad (2.27)$$

$$\cos(x \pm y) = (\cos x)(\cos y) \mp (\sin x)(\sin y) \qquad (2.28)$$

$$\tan(x + y) = \frac{(\tan x) + (\tan y)}{1 - (\tan x)(\tan y)} \qquad (2.29)$$

$$\sin(2x) = 2(\sin x)(\cos x) \qquad (2.30)$$

$$\cos(2x) = \cos^2 x - \sin^2 x \qquad (2.31)$$

$$\tan(2x) = \frac{2(\tan x)}{1 - \tan^2 x} \qquad (2.32)$$

$$(\sin x)(\sin y) = \frac{1}{2}[-\cos(x+y) + \cos(x-y)] \qquad (2.33)$$

$$(\cos x)(\cos y) = \frac{1}{2}[\cos(x+y) + \cos(x-y)] \qquad (2.34)$$

$$(\sin x)(\cos y) = \frac{1}{2}[\sin(x+y) + \sin(x-y)] \qquad (2.35)$$

$$(\sin x) + (\sin y) = 2 \sin\frac{x+y}{2} \cos\frac{x-y}{2} \qquad (2.36)$$

$$(\sin x) - (\sin y) = 2 \sin \frac{x-y}{2} \cos \frac{x+y}{2} \qquad (2.37)$$

$$(\cos x) + (\cos y) = 2 \cos \frac{x+y}{2} \cos \frac{x-y}{2} \qquad (2.38)$$

$$(\cos x) - (\cos y) = -2 \sin \frac{x+y}{2} \sin \frac{x-y}{2} \qquad (2.39)$$

$$A \cos(\omega t + \psi) + B \cos(\omega t + \phi) = C \cos(\omega t + \theta) \qquad (2.40)$$

where $C = [A^2 + B^2 - 2AB \cos(\phi - \psi)]^{1/2}$

$$\theta = \tan^{-1}\left(\frac{A \sin \psi + B \sin \phi}{A \cos \psi + B \cos \phi}\right)$$

$$A \cos(\omega t + \psi) + B \sin(\omega t + \phi) = C \cos(\omega t + \theta) \qquad (2.41)$$

where $C = [A^2 + B^2 - 2AB \sin(\phi - \psi)]^{1/2}$

$$\theta = \tan^{-1}\left(\frac{A \sin \psi - B \cos \phi}{A \cos \psi + B \sin \phi}\right)$$

Euler's identities

The following four equations, called *Euler's identities*, relate sinusoids and complex exponentials.

$$e^{jx} = \cos x + j \sin x \qquad (2.42)$$

$$e^{-jx} = \cos x - j \sin x \qquad (2.43)$$

$$\cos x = \frac{e^{jx} + e^{-jx}}{2} \qquad (2.44)$$

$$\sin x = \frac{e^{jx} - e^{-jx}}{2j} \qquad (2.45)$$

Series and product expansions [Ref. 9]

Listed below are infinite series expansions for the various trigonometric functions

$$\sin x = \sum_{n=0}^{\infty} \frac{(-1)^n x^{2n+1}}{(2n+1)!} \qquad (2.46)$$

$$\cos x = \sum_{n=0}^{\infty} \frac{(-1)^n x^{2n}}{(2n)!} \qquad (2.47)$$

$$\tan x = \sum_{n=1}^{\infty} \frac{(-1)^{n-1}2^{2n}(2^{2n}-1)B_{2n}x^{2n-1}}{(2n)!} \qquad |x| < \frac{\pi}{2} \qquad (2.48)$$

$$\cot x = \sum_{n=0}^{\infty} \frac{(-1)^{n}2^{2n}B_{2n}x^{2n-1}}{(2n)!} \qquad |x| < \pi \qquad (2.49)$$

$$\sec x = \sum_{n=0}^{\infty} \frac{(-1)^{n}E_{2n}x^{2n}}{(2n)!} \qquad |x| < \frac{\pi}{2} \qquad (2.50)$$

$$\csc x = \sum_{n=0}^{\infty} \frac{(-1)^{n-1}2(2^{2n-1}-1)B_{2n}x^{2n-1}}{(2n)!} \qquad |x| < \pi \qquad (2.51)$$

Values for the Bernoulli number B_n and Euler number E_n are listed in Tables 2.2 and 2.3, respectively. In some instances, the infinite product expansions for sine and cosine may be more convenient than the series expansions.

$$\sin x = \prod_{n=1}^{\infty} \left(1 - \frac{x^2}{n^2\pi^2}\right) \qquad (2.52)$$

$$\cos x = \prod_{n=1}^{\infty} \left(1 - \frac{4x^2}{(2n-1)^2\pi^2}\right) \qquad (2.53)$$

Orthonormality of sine and cosine

Two functions $\phi_1(t)$ and $\phi_2(t)$ are said to form an orthogonal set over the interval $[0, T]$ if

$$\int_0^T \phi_1(t)\phi_2(t)\, dt = 0 \qquad (2.54)$$

TABLE 2.2 Bernoulli Numbers ($B_n = N/D$ and $B_n = 0$ for $n = 3, 5, 7, \ldots$)

n	N	D
0	1	1
1	−1	2
2	1	6
4	−1	30
6	1	42
8	−1	30
10	5	66
12	−691	2730
14	7	6
16	−3,617	510
18	43,867	798
20	−174,611	330

TABLE 2.3 Euler Numbers ($E_n = 0$ for $n = 1, 3, 5, 7, \ldots$)

n	E_n
0	1
2	−1
4	5
6	−61
8	1,385
10	−50,521
12	2,702,765
14	−199,360,981
16	19,391,512,145
18	−2,404,879,675,441
20	370,371,188,237,525

The functions $\phi_1(t)$ and $\phi_2(t)$ are said to form an orthonormal set over the interval $[0, T]$ if, in addition to satisfying (2.54), each function has unit energy over the interval

$$\int_0^T [\phi_1(t)]^2 \, dt = \int_0^T [\phi_2(t)]^2 \, dt = 1 \tag{2.55}$$

Consider the two signals given by

$$\phi_1(t) = A \sin(\omega_0 t) \tag{2.56}$$

$$\phi_2(t) = A \cos(\omega_0 t) \tag{2.57}$$

The signals ϕ_1 and ϕ_2 will form an orthogonal set over the interval $[0, T]$ if $\omega_0 T$ is an integer multiple of π. The set will be orthonormal as well as orthogonal if $A^2 = 2/T$. The signals ϕ_1 and ϕ_2 will form an approximately orthonormal set over the interval $[0, T]$ if $\omega_0 T \gg 1$ and $A^2 = 2/T$. The orthonormality of sine and cosine can be derived as follows:

Substitution of (2.56) and (2.57) into (2.54) yields

$$\int_0^T \phi_1(t)\phi_2(t) \, dt = A^2 \int_0^T \sin \omega_0 t \cos \omega_0 t \, dt$$

$$= \frac{\theta^2}{2} \int_0^T [\sin(\omega_0 t + \omega_0 t) + \sin(\omega_0 t - \omega_0 t)] \, dt$$

$$= \frac{\theta^2}{2} \int_0^T \sin 2\omega_0 t \, dt = \frac{\theta^2}{2} \left(\frac{\cos 2\omega_0 t}{2\omega_0} \right) \Big|_{t=0}^T$$

$$= \frac{A^2}{4\omega_0 T}(1 - \cos 2\omega_0 T) \tag{2.58}$$

Thus if $\omega_0 T$ is an integer multiple of π, then $\cos(2\omega_0 T) = 1$, and ϕ_1 and ϕ_2 will be orthogonal. If $\omega_0 T \gg 1$, then (2.58) will be very small and reasonably approximated by zero; thus ϕ_1 and ϕ_2 can be considered as approximately orthogonal. The energy of $\phi_1(t)$ on the interval $[0, T]$ is given by

$$E_1 = \int_0^T [\phi_1(t)]^2 \, dt = A^2 \int_0^T \sin^2 \omega_0 t \, dt$$

$$= A^2 \left(\frac{t}{2} - \frac{\sin 2\omega_0 t}{4\omega_0} \right) \Big|_{t=0}^T$$

$$= A^2 \left(\frac{T}{2} - \frac{\sin 2\omega_0 T}{4\omega_0} \right) \tag{2.59}$$

For ϕ_1 to have unit energy, A^2 must satsify

$$A^2 = \left(\frac{T}{2} - \frac{\sin 2\omega_0 T}{4\omega_0}\right)^{-1} \tag{2.60}$$

When $\omega_0 T = n\pi$, then $\sin 2\omega_0 T = 0$. Thus (2.60) reduces to

$$A = \sqrt{\frac{2}{T}} \tag{2.61}$$

Substituting (2.61) into (2.59) yields

$$E_1 = 1 - \frac{\sin 2\omega_0 T}{2\omega_0 T} \tag{2.62}$$

When $\omega_0 T \gg 1$, the second term of (2.62) will be very small and reasonably approximated by zero, thus indicating that ϕ_1 and ϕ_2 are approximately orthonormal. In a similar manner, the energy of $\phi_1(t)$ can be found to be

$$E_2 = A^2 \int_0^T \cos^2 \omega_0 t \, dt$$

$$= A^2 \left(\frac{T}{2} + \frac{\sin 2\omega_0 T}{4\omega_0}\right) \tag{2.63}$$

Thus $\quad E_2 = 1 \quad$ if $A = \sqrt{\frac{2}{T}}$ and $\omega_0 T = n\pi$

$\quad\quad\quad E_2 \doteq 1 \quad$ if $A = \sqrt{\frac{2}{T}}$ and $\omega_0 T \gg 1$

2.3 Complex Numbers

A complex number z has the form $a + bj$, where a and b are real and $j = \sqrt{-1}$. The *real part* of z is a, and the *imaginary part* of z is b. Mathematicians use i to denote $\sqrt{-1}$, but electrical engineers use j to avoid confusion with the traditional use of i for denoting current. For convenience, $a + bj$ is sometimes represented by the ordered pair (a, b). The *modulus*, or *absolute value*, of z is denoted as $|z|$ and defined by

$$|z| = |a + bj| = \sqrt{a^2 + b^2} \tag{2.64}$$

The *complex conjugate* of z is denoted as z^* and is defined by

$$(z = a + bj) \Leftrightarrow (z^* = a - bj) \tag{2.65}$$

Conjugation distributes over addition, multiplication, and division:

$$(z_1 + z_2)^* = z_1^* + z_2^* \tag{2.66}$$

$$(z_1 z_2)^* = z_1^* z_2^* \tag{2.67}$$

$$\left(\frac{z_1}{z_2}\right)^* = \frac{z_1^*}{z_2^*} \tag{2.68}$$

Operations on complex numbers in rectangular form

Consider two complex numbers:

$$z_1 = a + bj \qquad z_2 = c + dj$$

The four basic arithmetic operations are then defined as:

$$z_1 + z_2 = (a + c) + j(b + d) \tag{2.69}$$

$$z_1 - z_2 = (a - c) + j(b - d) \tag{2.70}$$

$$z_1 z_2 = (ac \mp bd) + j(ad + bc) \tag{2.71}$$

see Ref Data p. 46-7
or Algebra 2 p. 42

$$\frac{z_1}{z_2} = \frac{ac + bd}{c^2 + d^2} + j\frac{bc - ad}{c^2 + d^2} \tag{2.72}$$

$\frac{1}{A + jB} = \frac{A}{A^2 + B^2} - j\frac{B}{A^2 + B^2}$

Polar form of complex numbers

A complex number of the form $a + bj$ can be represented by a point in a coordinate plane as shown in Fig. 2.2. Such a representation is called an *Argand diagram* [Ref. 1] in honor of Jean Robert Argand (1768–1822) who published a description of this graphical representation of complex numbers in 1806 [Ref. 2]. The point representing $a + bj$ can also be located using an angle θ and radius r as shown.

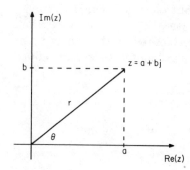

Figure 2.2 Argand diagram representation of a complex number.

From the definitions of sine and cosine given in (2.9) and (2.10), it follows that

$$a = r \cos \theta \qquad b = r \sin \theta$$

Therefore $\quad z = r \cos \theta + jr \sin \theta = r(\cos \theta + j \sin \theta) \qquad (2.73)$

The quantity $(\cos \theta + j \sin \theta)$ is sometimes denoted as cis θ. Making use of (2.41), we can rewrite (2.73) as:

$$z = r \text{ cis } \theta = r \exp(j\theta) \qquad (2.74)$$

The form in (2.74) is called the *polar form* of the complex number z.

Operations on complex numbers in polar form

Consider three complex numbers:

$$z = r(\cos \theta + j \sin \theta) = r \exp(j\theta)$$

$$z_1 = r_1(\cos \theta_1 + j \sin \theta_1) = r_1 \exp(j\theta_1)$$

$$z_2 = r_2(\cos \theta_2 + j \sin \theta_2) = r_2 \exp(j\theta_2)$$

Several operations can be conveniently performed directly upon complex numbers that are in polar form:

Multiplication:

$$\begin{aligned} z_1 z_2 &= r_1 r_2 [\cos(\theta_1 + \theta_2) + j \sin(\theta_1 + \theta_2)] \\ &= r_1 r_2 \exp[j(\theta_1 + \theta_2)] \end{aligned} \qquad (2.75)$$

Division:

$$\begin{aligned} \frac{z_1}{z_2} &= \frac{r_1}{r_2} [\cos(\theta_1 - \theta_2) + j \sin(\theta_1 - \theta_2)] \\ &= \frac{r_1}{r_2} \exp[j(\theta_1 - \theta_2)] \end{aligned} \qquad (2.76)$$

Powers:

$$\begin{aligned} z^n &= r^n [\cos(n\theta) + j \sin(n\theta)] \\ &= r^n \exp(jn\theta) \end{aligned} \qquad (2.77)$$

Roots:

$$\sqrt[n]{z} = z^{1/n} = r^{1/n}\left[\cos\left(\frac{\theta + 2k\pi}{n}\right) + j\sin\left(\frac{\theta + 2k\pi}{n}\right)\right]$$

$$= r^{1/n}\exp\frac{j(\theta + 2k\pi)}{n} \qquad k = 0, 1, 2, \ldots \qquad (2.78)$$

Equation (2.77) is known as *Demoivre's theorem*. In 1730, an equation similar to (2.78) was published by Abraham Demoivre (1667–1754) in his *Miscellanea analytica* [Ref. 2]. In Eq. (2.78), for a fixed n as k increases, the sinusoidal functions will take on only n distinct values. Thus there are n different nth roots of any complex number.

Logarithms of complex numbers

For the complex number $z = r\exp(j\theta)$, the natural logarithm of z is given by

$$\begin{aligned}\ln z &= \ln[r\exp(j\theta)] \\ &= \ln\{r\exp[j(\theta + 2k\pi)]\} \\ &= (\ln r) + j(\theta + 2k\pi) \qquad k = 0, 1, 2, \ldots\end{aligned} \qquad (2.79)$$

The *principal value* is obtained when $k = 0$.

2.4 Derivatives

Listed below are some derivative forms which often prove useful in theoretical analysis of communication systems.

$$\frac{d}{dx}\sin u = \cos u \frac{du}{dx} \qquad (2.80)$$

$$\frac{d}{dx}\cos u = -\sin u \frac{du}{dx} \qquad (2.81)$$

$$\frac{d}{dx}\tan u = \sec^2 u \frac{du}{dx} = \frac{1}{\cos^2 u}\frac{du}{dx} \qquad (2.82)$$

$$\frac{d}{dx}\cot u = \csc^2 u \frac{du}{dx} = \frac{1}{\sin^2 u}\frac{du}{dx} \qquad (2.83)$$

$$\frac{d}{dx}\sec u = \sec u \tan u \frac{du}{dx} = \frac{\sin u}{\cos^2 u}\frac{du}{dx} \qquad (2.84)$$

$$\frac{d}{dx}\csc u = -\csc u \cot u \frac{du}{dx} = \frac{-\cos u}{\sin^2 u}\frac{du}{dx} \qquad (2.85)$$

$$\frac{d}{dx}e^u = e^u \frac{du}{dx} \qquad (2.86)$$

$$\frac{d}{dx}\ln u = \frac{1}{u}\frac{du}{dx} \qquad (2.87)$$

$$\frac{d}{dx}\log u = \frac{\log e}{u}\frac{du}{dx} \qquad (2.88)$$

$$\frac{d}{dx}\left(\frac{u}{v}\right) = \frac{1}{v^2}\left(v\frac{du}{dx} - u\frac{dv}{dx}\right) \qquad (2.89)$$

Derivatives of polynomial ratios. Consider a ratio of polynomials given by

$$C(s) = \frac{A(s)}{B(s)} \qquad B(s) \neq 0 \qquad (2.90)$$

The derivative of $C(s)$ can be obtained using Eq. (2.89) to obtain

$$\frac{d}{ds}C(s) = [B(s)]^{-1}\frac{d}{ds}A(s) - A(s)[B(s)]^{-2}\frac{d}{ds}B(s) \qquad (2.91)$$

Equation (2.91) will be very useful in the application of the Heaviside expansion which is discussed in Sec. 7.6.

2.5 Integration

Large integral tables fill entire volumes and contain thousands of entries. However, a relatively small number of integral forms appear over and over again in the study of communications, and these are listed below.

$$\int \frac{1}{x} dx = \ln x \qquad (2.92)$$

$$\int e^{ax} dx = \frac{1}{a}e^{ax} \qquad (2.93)$$

$$\int xe^{ax} dx = \frac{ax-1}{a^2}e^{ax} \qquad (2.94)$$

$$\int \sin(ax) dx = -\frac{1}{a}\cos(ax) \qquad (2.95)$$

$$\int \cos(ax)\,dx = \frac{1}{a}\sin(ax) \tag{2.96}$$

$$\int \sin(ax+b)\,dx = -\frac{1}{a}\cos(ax+b) \tag{2.97}$$

$$\int \cos(ax+b)\,dx = \frac{1}{a}\sin(ax+b) \tag{2.98}$$

$$\int x\sin(ax)\,dx = -\frac{x}{a}\cos(ax) + \frac{1}{a^2}\sin(ax) \tag{2.99}$$

$$\int x\cos(ax)\,dx = \frac{x}{a}\sin(ax) + \frac{1}{a^2}\cos(ax) \tag{2.100}$$

$$\int \sin^2 ax\,dx = \frac{x}{2} - \frac{\sin 2ax}{4a} \tag{2.101}$$

$$\int \cos^2 ax\,dx = \frac{x}{2} + \frac{\sin 2ax}{4a} \tag{2.102}$$

$$\int x^2\sin ax\,dx = \frac{1}{a^3}(2ax\sin ax + 2\cos ax - a^2x^2\cos ax) \tag{2.103}$$

$$\int x^2\cos ax\,dx = \frac{1}{a^3}(2ax\cos ax - 2\sin ax + a^2x^2\sin ax) \tag{2.104}$$

$$\int \sin^3 x\,dx = -\frac{1}{3}\cos x(\sin^2 x + 2) \tag{2.105}$$

$$\int \cos^3 x\,dx = \frac{1}{3}\sin x(\cos^2 x + 2) \tag{2.106}$$

$$\int \sin x\cos x\,dx = \frac{1}{2}\sin^2 x \tag{2.107}$$

$$\int \sin(mx)\cos(nx)\,dx = \frac{-\cos(m-n)x}{2(m-n)} - \frac{\cos(m+n)x}{2(m+n)} \quad (m^2 \neq n^2) \tag{2.108}$$

$$\int \sin^2 x\cos^2 x\,dx = \frac{1}{8}\left[x - \frac{1}{4}\sin(4x)\right] \tag{2.109}$$

$$\int \sin x\cos^m x\,dx = \frac{-\cos^{m+1} x}{m+1} \tag{2.110}$$

$$\int \sin^m x \cos x \, dx = \frac{\sin^{m+1} x}{m+1} \qquad (2.111)$$

$$\int \cos^m x \sin^n x \, dx = \frac{\cos^{m-1} x \sin^{n+1} x}{m+n}$$
$$+ \frac{m-1}{m+n} \int \cos^{m-2} x \sin^n x \, dx \qquad (m \neq -n) \quad (2.112)$$

$$\int \cos^m x \sin^n x \, dx = \frac{-\cos^{m+1} x \sin^{n-1} x}{m+n}$$
$$+ \frac{n-1}{m+n} \int \cos^m x \sin^{n-2} x \, dx \qquad (m \neq -n) \quad (2.113)$$

$$\int u \, dv = uv - \int v \, du \qquad (2.114)$$

2.6 Dirac Delta Function

In all of electrical engineering, there is perhaps nothing which is responsible for more hand waving than is the so-called *delta function*, or *impulse function*, which is denoted $\delta(t)$ and which is usually depicted as a vertical arrow at the origin as shown in Fig. 2.3. This function is often called the *Dirac delta function* in honor of Paul Dirac (1902–1984), an English physicist who used delta functions extensively in his work on quantum mechanics. A number of nonrigorous approaches for defining the impulse function can be found throughout the literature. A *unit impulse* is often loosely described as having a zero width and an infinite amplitude at the origin such that the total area under the impulse is equal to unity. How is it possible to claim that zero times infinity equals 1? The trick involves defining

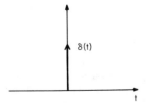

Figure 2.3 Graphical representation of the Dirac delta function.

a sequence of functions $f_n(t)$ such that

$$\int_{-\infty}^{\infty} f_n(t)\, dt = 1 \tag{2.115}$$

and
$$\lim_{n \to \infty} f_n(t) = 0 \quad \text{for } t \neq 0 \tag{2.116}$$

The delta function is then defined as:

$$\delta(t) = \lim_{n \to \infty} f_n(t) \tag{2.117}$$

Example Let a sequence of pulse functions $f_n(t)$ be defined as:

$$f_n(t) = \begin{cases} \dfrac{n}{2} & |t| \leq 1/n \\ 0 & \text{otherwise} \end{cases} \tag{2.118}$$

Equation (2.115) is satisfied since the area of the pulse is equal to $(2n)(n/2) = 1$ for all n. The pulse width decreases and the pulse amplitude increases as n approaches infinity. Therefore we intuitively sense that this sequence must also satisfy (2.116). Thus the impulse function can be defined as the limit of (2.118) as n approaches infinity. Using similar arguments, it can be shown that the impulse can also be defined as the limit of a sequence of sinc functions or gaussian pulse functions.

A second approach entails simply defining $\delta(t)$ to be that function which satisfies

$$\int_{-\infty}^{\infty} \delta(t)\, dt = 1 \quad \text{and} \quad \delta(t) = 0 \quad \text{for } t \neq 0 \tag{2.119}$$

In a third approach, $\delta(t)$ is defined as that function which exhibits the property

$$\int_{-\infty}^{\infty} \delta(t) f(t)\, dt = f(0) \tag{2.120}$$

While any of these three approaches is adequate to introduce the delta function into an engineer's repertoire of analytical tools, none of the three is sufficiently rigorous to satisfy mathematicians or discerning theoreticians. In particular, notice that none of the approaches presented deals with the thorny issue of just what the value of $\delta(t)$ is for $t = 0$. The rigorous definition of $\delta(t)$, introduced in 1950 by Laurent Schwartz (1915–) in Ref. 8, rejects the notion that the impulse is an ordinary function and instead defines it as a *distribution*.

Distributions

Let S be the set of functions $f(x)$ for which the nth derivative $f^{[n]}(x)$ exists for any n and all x. Furthermore, each $f(x)$ decreases sufficiently fast at infinity such that

$$\lim_{x \to \infty} x^n f(x) = 0 \quad \text{for all } n \tag{2.121}$$

A distribution, often denoted $\phi(x)$, is defined as a continuous linear mapping from the set S to the set of complex numbers. Notationally, this mapping is represented as an inner product:

$$\int_{-\infty}^{\infty} \phi(x) f(x) \, dx = z \tag{2.122}$$

or alternatively as:

$$\langle \phi(x), f(x) \rangle = z \tag{2.123}$$

Notice that no claim is made that ϕ is a function capable of mapping values of x into corresponding values $\phi(x)$. In some texts (such as Ref. 4), $\phi(x)$ is referred to as a *functional* or a *generalized function*. The distribution ϕ is defined only through the impact that it has upon other functions. The impulse function is a distribution defined by the following:

$$\int_{-\infty}^{\infty} \delta(t) f(t) \, dt = f(0) \tag{2.124}$$

The equation (2.124) looks exactly like (2.120), but defining $\delta(t)$ as a distribution eliminates the need to tap-dance around the issue of assigning a value to $\delta(0)$. Furthermore, the impulse function is elevated to a more substantial foundation from which several useful properties may be rigorously derived. For a more in-depth discussion of distributions other than $\delta(t)$, the interested reader is referred to Chap. 4 of Ref. 5.

Properties of the delta distribution

It has been shown [Eqs. (2.118) through (2.121)] that the delta distribution exhibits the following properties:

$$\int_{-\infty}^{\infty} \delta(t) \, dt = 1 \tag{2.125}$$

$$\frac{d}{dt} \delta(t) = \lim_{\tau \to 0} \frac{\delta(t) - \delta(t - \tau)}{\tau} \tag{2.126}$$

$$\int_{-\infty}^{\infty} \delta(t - t_0) f(t) \, dt = f(t_0) \qquad (2.127)$$

$$\delta(at) = \frac{1}{|a|} \delta(t) \qquad (2.128)$$

$$\delta(t_0) f(t) = f(t_0) \delta(t_0) \qquad (2.129)$$

$$\delta_1(t - t_1) * \delta_2(t - t_2) = \delta[t - (t_1 + t_2)] \qquad (2.130)$$

In Eq. (2.129) $f(t)$ is an ordinary function that is continuous at $t = t_0$, and in Eq. (2.130) the asterisk denotes convolution.

2.7 Bessel Functions

Bessel functions of the first kind

The *Bessel functions of the first kind of order n and argument x* are usually denoted by $J_n(x)$ and are defined by

$$J_n(x) = \frac{1}{2\pi} \int_{-\pi}^{\pi} \exp[jx \sin(\theta) - jn\theta] \, d\theta \qquad (2.131a)$$

$$= \frac{1}{\pi} \int_{0}^{\pi} \cos[x \sin(\theta) - n\theta] \, d\theta \qquad (2.131b)$$

Bessel functions as defined by (2.131a and b) are used in the analysis of FM signals.

Bessel function identities

Listed below are some identities that involve J_n:

$$J_n(x) = \sum_{m=0}^{\infty} \frac{(-1)^m (x/2)^{2m+n}}{m!(n+m)!} \qquad (2.132)$$

$$J_n(-x) = J_{-n}(x) = (-1)^n J_n(x) \qquad (2.133)$$

$$J_{n+1}(x) = \frac{2n}{x} J_n(x) - J_{n-1}(x) \qquad (2.134)$$

$$J_n'(x) = \frac{1}{2}[J_{n-1}(x) - J_{n+1}(x)] \qquad (2.135)$$

$$J_n''(x) = \frac{1}{4}[J_{n-2}(x) - 2J_n(x) + J_{n+2}(x)] \qquad (2.136)$$

$$J_n^{(k)}(x) = 2^{-k} \sum_{m=0}^{\infty} \frac{(-1)^m k!}{m!(k-m)!} J_{n-k+2m}(x) \qquad (2.137)$$

(Note: $J_n^{(k)}$ denotes the kth derivative of J_n.)

$$\frac{d^k}{dx^k}[x^n J_n(x)] = x^n J_{n-k}(x) \tag{2.138}$$

$$\frac{d^k}{dx^k}\left[\frac{J_n(x)}{x^n}\right] = (-1)^k \frac{J_{n+k}(x)}{x^n} \tag{2.139}$$

$$\exp(jx\sin\theta) = \sum_{n=-\infty}^{\infty} J_n(x)\exp(jn\theta) \tag{2.140}$$

$$\exp(j\phi + jx\sin\theta) = \sum_{n=-\infty}^{\infty} J_n(x)\exp(j\phi + jn\theta) \tag{2.141}$$

$$\exp(jx\cos\theta) = \sum_{n=-\infty}^{\infty} j^n J_n(x)\exp(jn\theta) \tag{2.142}$$

$$\cos x = J_0(x) + 2\sum_{n=1}^{\infty} (-1)^n J_{2n}(x) \tag{2.143}$$

$$\sin x = 2\sum_{n=0}^{\infty} (-1)^n J_{2n+1}(x) \tag{2.144}$$

$$\cos(x\sin\theta) = J_0(x) + 2\sum_{n=1}^{\infty} J_{2n}(x)\cos(2n\theta) \tag{2.145}$$

$$\sin(x\sin\theta) = 2\sum_{n=1}^{\infty} J_{2n-1}(x)\sin[(2n-1)\theta] \tag{2.146}$$

$$\cos(x\cos\theta) = J_0(x) + 2\sum_{n=1}^{\infty} (-1)^n J_{2n}(x)\cos(2n\theta) \tag{2.147}$$

$$\sin(x\cos\theta) = 2\sum_{n=1}^{\infty} (-1)^n J_{2n+1}(x)\cos[(2n+1)\theta] \tag{2.148}$$

$$\sin(\phi + x\sin\theta) = \sum_{n=-\infty}^{\infty} J_n(x)\sin(\phi + n\theta) \tag{2.149}$$

$$\cos(\phi + x\sin\theta) = \sum_{n=-\infty}^{\infty} J_n(x)\cos(\phi + n\theta) \tag{2.150}$$

$$\sin(\phi + x\cos\theta) = \sum_{n=-\infty}^{\infty} J_n(x)\sin\left(\phi + n\theta + \frac{n\pi}{2}\right) \tag{2.151}$$

$$\cos(\phi + x\cos\theta) = \sum_{n=-\infty}^{\infty} J_n(x)\cos\left(\phi + n\theta + \frac{n\pi}{2}\right) \tag{2.152}$$

$$J_0(x) = \frac{1}{\pi} \int_0^\pi \cos(x \sin \theta) \, d\theta \qquad (2.153)$$

$$J_0(x) = \frac{1}{\pi} \int_0^\pi \cos(x \cos \theta) \, d\theta \qquad (2.154)$$

$$J_0(x) = \frac{2}{\pi} \int_0^\infty \sin(x \cosh \theta) \, d\theta \qquad (x \text{ positive and real}) \qquad (2.155)$$

Modified Bessel functions of the first kind

The *modified Bessel function of the first kind* of order n is usually denoted by $I_n(x)$ and is defined as:

$$I_n(x) = \frac{1}{2\pi} \int_{-\pi}^\pi \exp(x \cos \theta) \cos(n\theta) \, d\theta \qquad (2.156)$$

The modified Bessel function of the first kind of order zero is used in the analysis of Rice random variables (see Sec. 4.8).

Identities for modified Bessel functions

Listed below are some identities that involve I_n.

$$I_n(x) = \sum_{m=0}^\infty \frac{(x/2)^{2m+n}}{m!(n+m)!} \qquad (2.157)$$

$$I_{-n}(x) = I_n(x) \qquad (2.158)$$

$$I_n(-x) = (-1)^n I_n(x) \qquad (2.159)$$

$$\exp(x \cos \theta) = \sum_{n=-\infty}^\infty I_n(x) \exp(jn\theta) \qquad (2.160)$$

$$\exp(x \cos \theta) = I_0(x) + 2 \sum_{n=1}^\infty I_n(x) \cos(n\theta) \qquad (2.161)$$

$$\frac{d}{dx}[x^n I_n(x)] = x^n I_{n-1}(x) \qquad (2.162)$$

$$\frac{d}{dx}\left[\frac{I_n(x)}{x^n}\right] = \frac{I_{n+1}(x)}{x^n} \qquad (2.163)$$

$$I_0(x) = \frac{1}{\pi} \int_0^\pi \exp(x \cos \theta) \, d\theta \qquad (2.164)$$

$$I_0(x) = \frac{1}{\pi} \int_0^\pi \cosh(x \cos \theta) \, d\theta \qquad (2.165)$$

Evaluation of Bessel functions

For large values of x, $I_0(x)$ can be approximated as:

$$I_0(x) \approx \frac{\exp(x)}{\sqrt{2\pi x}} \qquad (2.166)$$

For small values of x, $I_0(x)$ is approximately equal to 1.

2.8 Special Functions

Gamma function

The *gamma function of* x, denoted $\Gamma(x)$, is defined by

$$\Gamma(x) \triangleq \int_0^\infty t^{x-1} e^{-t} \, dt \qquad n > 0 \qquad (2.167)$$

For integer values of x, the gamma function and factorial are related via

$$\Gamma(n+1) = n! \qquad (2.168)$$

In some texts, Eq. (2.168) is used to extend the definition of factorial to include noninteger values of n. As a consequence of (2.168), we can immediately establish the following recursion formulas.

$$\Gamma(x+1) = x\Gamma(x) \qquad (2.169)$$

$$\Gamma(x) = \frac{\Gamma(x+1)}{x} \qquad (2.170)$$

Equation (2.170) can be used to "work backward" to define the gamma function for negative values of x. Some other useful gamma function formulas include

$$\Gamma\left(\frac{1}{2}\right) = \sqrt{\pi} \qquad (2.171)$$

$$\Gamma\left(\frac{1}{2} + m\right) = \frac{1 \cdot 3 \cdot 5 \cdot \ldots \cdot (2m-1)}{2^m} \sqrt{\pi} \qquad m = 1, 2, 3, \ldots \qquad (2.172)$$

$$\Gamma\left(\frac{1}{2} - m\right) = \frac{(-2)^m \sqrt{\pi}}{1 \cdot 3 \cdot 5 \cdot \ldots \cdot (2m-1)} \qquad m = 1, 2, 3, \ldots \qquad (2.173)$$

$$\Gamma(X)\Gamma(1-x) = \frac{\pi}{\sin x\pi} \qquad (2.174)$$

$$\frac{\Gamma(2x)}{2^{2x-1}} = \frac{\Gamma(x)\Gamma[x + (1/2)]}{\sqrt{\pi}} \qquad (2.175)$$

Confluent hypergeometric function

The confluent hypergeometric function (chf), denoted $_1F_1(\alpha; \beta; x)$, is defined as:

$$_1F_1(\alpha; \beta; x) = \sum_{n=0}^{\infty} \frac{\Gamma(\alpha + n)\Gamma(\beta)x^n}{\Gamma(\alpha)\Gamma(\beta + n)n!} \qquad (2.176)$$

The following identities, called *Kummer's transformations*, are sometimes useful in manipulating expressions involving the chf.

$$_1F_1(\alpha; \beta; x) = e^x \,_1F_1(\beta - \alpha; \beta; -x) \qquad (2.177)$$

$$_1F_1(\alpha; \beta; -x) = e^{-x} \,_1F_1(\beta - \alpha; \beta; x) \qquad (2.178)$$

The chf is used in computing the moments of Rice distributions.

Error function

The *error function* of x, written as erf x, is defined by

$$\operatorname{erf} x \triangleq \frac{2}{\sqrt{\pi}} \int_0^x \exp(-u^2)\, du \qquad (2.179)$$

The complementary error function of x, written as erfc x, is defined by

$$\operatorname{erfc} x \triangleq \frac{2}{\sqrt{\pi}} \int_x^{\infty} \exp(-u^2)\, du$$

$$= 1 - \operatorname{erf} x \qquad (2.180)$$

The integral in (2.179) cannot be solved in closed form, but numerically computed values have been extensively tabulated. A short table of values for erf x is presented in Table 2.4. Although erfc x cannot be evaluated in closed form, analytical expressions for upper and lower bounds have been established.

$$\operatorname{erfc} x > \left(1 - \frac{1}{2x^2}\right) \frac{\exp(-x^2)}{\sqrt{\pi x}} \qquad (2.181)$$

$$\operatorname{erfc} x < \frac{\exp(-x^2)}{\sqrt{\pi x}} \qquad (2.182)$$

For values of $x \geq 2$, both (2.181) and (2.182) closely approximate erfc x.

TABLE 2.4 Values of the Error Function

x	erf x	x	erf x
0.00	0.0000000	1.00	0.8427008
0.05	0.0563720	1.05	0.8624361
0.10	0.1124629	1.10	0.8802051
0.15	0.1679960	1.15	0.8961238
0.20	0.2227026	1.20	0.9103140
0.25	0.2763264	1.25	0.9229001
0.30	0.3286268	1.30	0.9340079
0.35	0.3793821	1.35	0.9437622
0.40	0.4283924	1.40	0.9522851
0.45	0.4754817	1.45	0.9596950
0.50	0.5204999	1.50	0.9661051
0.55	0.5633234	1.55	0.9716227
0.60	0.0638561	1.60	0.9763484
0.65	0.6420293	1.65	0.9803756
0.70	0.6778012	1.70	0.9837904
0.75	0.7111556	1.75	0.9866717
0.80	0.7421010	1.80	0.9890905
0.85	0.7706681	1.85	0.9911110
0.90	0.7969082	1.90	0.9927904
0.95	0.8208908	1.95	0.9941793
1.00	0.8427008	2.00	0.9953223

The following expansions may be useful in work involving the error function.

$$\int_0^x \text{erf}(y)\, dy = x\, \text{erf}(x) - \frac{1}{\sqrt{\pi}} [1 - \exp(-x^2)] \qquad (2.183)$$

$$\text{erf}(x) = \frac{2}{\sqrt{\pi}} \exp(-x^2) \sum_{n=0}^{\infty} \frac{2^{2n+1} x^{2n+1} (n+1)!}{(2n+2)!} \qquad (2.184)$$

$$\text{erf}(x) = \frac{2}{\sqrt{\pi}} \sum_{n=0}^{\infty} \frac{(-1)^n x^{2n+1}}{n!(2n+1)} \qquad (2.185)$$

Values of erf for negative x are obtained using the identity $\text{erf}(-x) = -\text{erf}(x)$.

Q Function

Closely related to the error function is the Q function, defined by

$$Q(x) = \frac{1}{\sqrt{2\pi}} \int_x^{\infty} \exp\left(\frac{-u^2}{2}\right) du \qquad (2.186)$$

Comparison of (2.179), (2.180), and (2.186) reveals that

$$Q(x) = \frac{1}{2}\operatorname{erfc}\left(\frac{x}{\sqrt{2}}\right) \qquad (2.187)$$

$$\operatorname{erfc} x = 2Q(\sqrt{2}x) \qquad (2.188)$$

While the Q function cannot be evaluated in closed form, analytical expressions for upper and lower bounds have been established.

$$Q(x) \geq \left(1 - \frac{1}{x^2}\right)\frac{\exp(-x^2/2)}{x\sqrt{2\pi}} \qquad (2.189)$$

$$Q(x) \leq \frac{\exp(-x^2/2)}{x\sqrt{2\pi}} \qquad (2.190)$$

For values of $x > 3$, both (2.189) and (2.190) closely approximate $Q(x)$. A short table of values for $Q(x)$ is presented in Table 2.5. Values of $Q(x)$ for negative x can be obtained by making use of the fact that $Q(-x) = 1 - Q(x)$. The Q function of a single argument should not be confused with the so-called Marcum Q function or generalized Q function which has two arguments.

TABLE 2.5 Values of the Q Function

x	$Q(x)$	x	$Q(x)$
0.0	0.500000	2.0	0.022750
0.1	0.460172	2.1	0.017864
0.2	0.420740	2.2	0.013903
0.3	0.382089	2.3	0.010724
0.4	0.344578	2.4	0.008198
0.5	0.308538	2.5	0.006210
0.6	0.274253	2.6	0.004661
0.7	0.241964	2.7	0.003467
0.8	0.211855	2.8	0.002555
0.9	0.184060	2.9	0.001866
1.0	0.158655	3.0	0.001350
1.1	0.135666	3.1	0.000968
1.2	0.115070	3.2	0.000687
1.3	0.096800	3.3	0.000483
1.4	0.080757	3.4	0.000337
1.5	0.066807	3.5	0.000233
1.6	0.054799	3.6	0.000159
1.7	0.044565	3.7	0.000108
1.8	0.035930	3.8	0.000072
1.9	0.028717	3.9	0.000048
2.0	0.022750	4.0	0.000032

Marcum Q function

The Marcum Q function $Q(a, b)$ is the probability that the envelope of a sine wave of amplitude a plus additive gaussian noise of unit variance exceeds some value b. The function $Q(a, b)$ equals the area under a Rice pdf lying to the right of the value b.

$$Q(a, b) = \int_b^\infty x I_0(ax) \exp\left(-\frac{x^2 + a^2}{2}\right) dx \qquad (2.191)$$

The Marcum Q function exhibits the following properties:

$$Q(a, 0) = 1 \qquad (2.192)$$

$$Q(0, b) = \exp\left(\frac{-b^2}{2}\right) \qquad (2.193)$$

Equation (2.192) is simply a restatement of the fact that the envelope is always positive. When $a = 0$ as in (2.193), the pdf of the envelope reduces to a Rayleigh distribution. The complement of the Q function given by

$$R(a, b) = 1 - Q(a, b)$$

is known as the *circular convergence function*.

2.9 Combinations and Permutations

A set of n elements has C subsets of k elements each, where C is given by

$$C = \binom{n}{k} \equiv \frac{n!}{k!(n-k)!} \qquad (2.194)$$

The coefficient C is sometimes called a *binomial coefficient* and equals the number of "combinations of n things taken k at a time."

Each group of k elements could be ordered in $k!$ different ways. Thus a set of n elements has P *ordered* subsets of k elements each, where P is given by

$$P = \frac{n!}{(n-k)!} \qquad (2.195)$$

The coefficient P is sometimes called "permutations of n things taken k at a time."

2.10 References

1. M. R. Spiegel: *Laplace Transforms*, Schaum's Outline Series, McGraw-Hill, New York, 1965.
2. C. B. Boyer: *A History of Mathematics*, Wiley, New York, 1968.
3. W. H. Press, B. P. Flannery, S. A. Teukolsky, and W. T. Vetterling: *Numerical Recipes*, Cambridge University Press, Cambridge, 1986.
4. A. Papoulis: *The Fourier Integral and Its Applications*, McGraw-Hill, New York, 1962.
5. H. J. Weaver: *Theory of Discrete and Continuous Fourier Analysis*, Wiley, New York, 1989.
6. E. O. Brigham: *The Fast Fourier Transform*, Prentice-Hall, Englewood Cliffs, N.J., 1974.
7. R. J. Schwartz and B. Friedland: *Linear Systems*, McGraw-Hill, New York, 1965.
8. L. Schwartz: *Thèorie des distributions*, Herman & Cie, Paris, 1950.
9. M. Abramowitz and I. A. Stegun: *Handbook of Mathematical Functions*, National Bureau of Standards, Appl. Math Series 55, 1966.

Chapter 3

Probability and Random Variables

3.1 Randomness and Probability

Consider the space $\mathscr{S}$ comprising a (possibly infinite) number of events. An example of such a space would be the set of all possible outcomes from rolling a single die. In this space there would be six events—one corresponding to each face of the die. The space $\mathscr{S}$ in its entirety can be called the *certain event* since it is certain that one of the included events must occur. (Let us assume that a die must land on one of its faces—we shall neglect the pathological cases that require the die to remain stably balanced on an edge or on a vertex.) Each event can also be called an *experimental outcome*. Actually, any subset of $\mathscr{S}$ is an event, with those subsets containing a single element being further distinguished as *elementary events*. Some texts such as Ref. 6 refer to elementary events as *sample points* of the experiment. In our example, a nonelementary event might be the rolling of an odd number, while an elementary event would be the rolling of a 5. The subset of $\mathscr{S}$, which is the empty set (denoted as Φ), is called the *impossible event*. A numeric value $P(A)$ can be assigned to every event A in the space $\mathscr{S}$. In probability theory this value is called the *probability of the event* A and is assigned such that

1. The probability of the certain event is unity

$$P(\mathscr{S}) = 1$$

2. All probabilities of events within $\mathscr{S}$ are nonnegative.

$$P(A) > 0$$

3. If events A and B share no common outcomes (that is, if A and B are *mutually exclusive*), then the probability of either A or B occurring is equal to the probability of event A occurring plus the probability of event B occurring.

$$[A \cap B = \Phi] \Rightarrow P(A + B) = P(A) + P(B)$$

A single execution of an experiment that selects an event from $\mathscr{S}$ is called a *trial*. A collection of events is called a *field* if it contains the certain event and is closed under finite union and complementation. A collection of events is called a *sigma field* (or σ field) if it contains the certain event and is closed under countable union and complementation. A σ field is also called a *Borel field* in honor of Emile Borel (1871–1956), a French mathematician who in 1909 introduced the concept of an ∞-distributed b-ary sequence (Ref. 5). Every σ field is also a field since closure under countable union implies closure under finite union.

Joint and conditional probabilities

The probability that both event A and event B occur is called the *joint probability* of A and B and is denoted as $P(AB)$ or $P(A \cap B)$. The *conditional probability* that event A will occur given that event B has occurred is denoted by $P(A \mid B)$ and is defined as:

$$P(A \mid B) = \frac{P(AB)}{P(B)} = \frac{P(A \cap B)}{P(B)} \qquad (3.1)$$

If B is a subset of A (that is, $B \subset A$), then

$$P(A \mid B) = \frac{P(AB)}{P(B)} = \frac{P(B)}{P(B)} = 1 \qquad (3.2)$$

If A is a subset of B (that is, $A \subset B$), then

$$P(A \mid B) = \frac{P(AB)}{P(B)} = \frac{P(A)}{P(B)} \geq P(A) \qquad (3.3)$$

since $P(B) \leq 1$. If events A and B are mutually exclusive [that is, $P(A \cap B) = 0$] and exhaustive [that is, $P(A \cup B) = 1$], then

$$P(A \mid B) = \frac{P(B \mid A)P(A)}{P(B)} \qquad (3.4)$$

The relationship expressed in (3.4) is called *Bayes rule*. The relationship expressed in (3.4) can be extended to the case of more than two events as follows: If the events A_n $(n = 1, 2, \ldots, N)$ are mutually exclusive and exhaustive, that is,

$$A_n \cap A_m = \Phi \quad m, n \in \{1, 2, \ldots, N\}$$
$$m \neq n$$

$$\bigcup_{n=1}^{N} A_n = \mathscr{S}$$

and if B is any event having nonzero probability, then

$$P(A_n \mid B) = \frac{P(A_n, B)}{P(B)}$$
$$= \frac{P(B \mid A_n)P(A_n)}{\sum_{m=1}^{N} P(B \mid A_m)P(A_m)} \quad (3.5)$$

Independent events

Two events A and B are called *independent*, or *statistically independent*, if $P(A \cap B) = P(A)P(B)$. Furthermore, if A and B are independent, then

$$P(A \mid B) = P(A) \quad \text{and} \quad P(B \mid A) = P(B)$$

For the N events $A_1, A_2, \ldots, A_N$ to be independent, their probabilities must satisfy

$$P(A_1 \cap A_2 \cap \cdots \cap A_N) = P(A_1)P(A_2) \cdots P(A_N) \quad (3.6)$$

Note that pairwise independence between each pair of events is not sufficient to establish that all N events are mutually independent.

3.2 Bernoulli Trials

In many applications a single experiment can have only two possible outcomes. An electronic component can be tested and found either defective or not defective. A coin can be tossed and land on either heads or tails. If n experiments are performed—if n coins are tossed or if a single coin is tossed n times—the probability that a particular outcome will be observed k times is given by

$$\binom{n}{k} p^k (1-p)^{n-k} = \frac{n!}{k!(n-k)!} p^k (1-p)^{n-k} \quad (3.7)$$

where p is the probability of the desired outcome in a single experiment. [The notation used in (3.7) is discussed in Sec. 2.9.]

Example Suppose that 10 percent of the resistors produced by a particular machine are defective. If we examine a sample of 20 resistors, what is the probability that less than 3 will be defective?

solution To obtain the probability of finding at most 2 defective units, we must sum together the probability of finding no defects, the probability of finding exactly 1 defective unit, and the probability of finding exactly 2 defective units.

$$P(0 \text{ defects}) = \binom{20}{0}(0.10)^0(0.90)^{20} = 0.12158$$

$$P(1 \text{ defect}) = \binom{20}{1}(0.10)^1(0.90)^{19} = 0.27017$$

$$P(2 \text{ defects}) = \binom{20}{2}(0.10)^2(0.90)^{18} = 0.28518$$

$$P(\text{less than 3 defects}) = 0.6769$$

3.3 Random Variables

If we conduct an experiment (such as rolling a die), we can assign a numeric value to each possible outcome of the experiment. The rule for assigning values to outcomes is called a *random variable*. Although it is called a random *variable*, it is not really a variable in the usual sense. It is more like a function which maps each experimental outcome in the domain into the corresponding numeric value in the range. The outcomes of rolling a single die can be denoted $f_1, f_2, \ldots, f_6$ as shown in Fig. 3.1. We can define a random variable $\mathbf{x}(f_i)$, which assigns a value to each outcome. The obvious choice would be for $f_1 = 1, f_2 = 2$, and so on. However, we are not limited to just this option—we can define the mapping in many different ways, such as:

$$\mathbf{x}(f_i) = 5i \qquad \mathbf{x}(f_i) = i^2 \qquad \mathbf{x}(f_i) = i - 1$$

Remember: f_i represents the possible outcomes of the experiment, while $\mathbf{x}(f_i)$ is the random variable which assigns numeric values to each of these outcomes. (This numeric value should not be confused with the numeric probability value that is assigned to each outcome. In the case of a fair die, the probability of each face landing up is $\frac{1}{6}$ regardless of what numeric score that may be printed on or assigned to the face.)

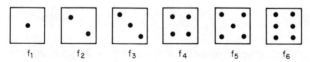

Figure 3.1 Possible outcomes of rolling a single die.

Cumulative distribution functions

The *cumulative distribution function* $F_x(x)$ associated with a random variable **x** yields the probability that **x** does not exceed x:

$$F_x(x) \equiv P\{\mathbf{x} \le x\}$$

Since $F_x(x)$ is a probability, obviously $0 \le F_x(x) \le 1$. Even though F_x is a function of x and not **x**, it is often referred to as the *distribution function of the random variable* **x**. In some of the literature, it is also called the *probability distribution function*, but this is usually avoided since the apparent abbreviation of pdf could also be interpreted as "probability density function." Although some authors use the lowercase pdf to denote "probability density function" and the uppercase PDF to denote "probability distribution function," most authors do not since this usage is potentially confusing.

In cases where the independent variable and the subscript on F differ, the subscript indicates the random variable (RV) of interest. Thus $F_x(y)$ represents the distribution function (evaluated at y) of the RV **x**. This is equal to the probability that the value of the RV **x** is less than or equal to y. In cases where the independent variable and the subscript are the same, the subscript is often omitted.

Properties of distribution functions

If $F(x)$ is the distribution function of the RV **x**, then it will exhibit the following properties:

$$F(-\infty) = 0 \qquad (3.8)$$

$$F(+\infty) = 1 \qquad (3.9)$$

$$\text{If } x_1 > x_2, \text{ then } F(x_1) \le F(x_2) \qquad (3.10)$$

$$\text{If } F(x_1) = 0, \text{ then } F(x) = 0 \quad \text{for all } x \le x_1 \qquad (3.11)$$

$$P\{\mathbf{x} > x\} = 1 - F(x) \qquad (3.12)$$

$$P\{x_1 < \mathbf{x} < x_2\} = F(x_2) - F(x_1) \qquad (3.13)$$

$$P\{\mathbf{x} = x\} = F(x) - \lim_{0 < \epsilon \to 0} F(x - \epsilon) \qquad (3.14)$$

$$P\{x_1 \le \mathbf{x} \le x_2\} = F(x_2) - \lim_{0 < \epsilon \to 0} F(x_1 - \epsilon) \qquad (3.15)$$

Probability density function

The derivative of the distribution function is called the *probability density function* (pdf), *density function*, or *frequency function* of the random variable **x**.

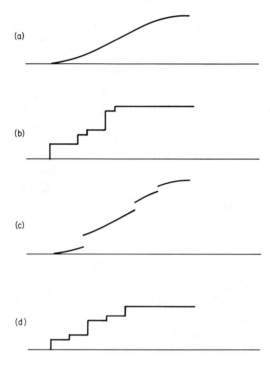

Figure 3.2 Distribution functions for different types of random variables. (a) Continuous. (b) Discrete. (c) Mixed-type. (d) Lattice-type.

Types of random variables

A *continuous* random variable has a continuous distribution function as shown in Fig. 3.2a. A *discrete* random variable has a stepped distribution function, as shown in Fig. 3.2b. A *mixed-type* random variable has a distribution function, such as shown in Fig. 3.2c, that is discontinuous but not necessarily flat between discontinuities. Sometimes the terminology *lattice-type* random variable is used to describe a discontinuous random variable whose cumulative distribution function has discontinuities at regularly spaced intervals, as shown in Fig. 3.2d.

3.4 Moments of a Random Variable

Mean

The *mean* $E(\mathbf{x})$ of a *discrete* random variable $\mathbf{x}$ is defined as:

$$E(\mathbf{x}) = \sum_{i=-\infty}^{\infty} x_i p_i \qquad (3.16)$$

where p_i is the probability of the event $\mathbf{x} = x_i$. The mean of a *continuous* random variable is defined as:

$$E(x) = \int_{-\infty}^{\infty} x f(x)\, dx \qquad (3.17)$$

where $f(x)$ is the probability density function of the RV **x**.

- The mean is also called the *expected value, expectation, ensemble average,* or *statistical average* and is typically denoted by an overbar or by η_x, η, or μ.

$$\bar{x} = \eta_x = \eta = \mu = E(\mathbf{x})$$

- The mean of the sum of two random variables is equal to the sum of their means.

$$E(\mathbf{x} + \mathbf{y}) = E(\mathbf{x}) + E(\mathbf{y}) \qquad (3.18)$$

- In general, the mean of a product of two RVs is not equal to the product of their individual means. However, the mean of the product *will* equal the product of the means if **x** and **y** are uncorrelated.

- The mean of a constant times a random variable is equal to the constant times the mean.

$$E(a\mathbf{x}) = aE(\mathbf{x}) \qquad (3.19)$$

- Taken together (3.18) and (3.19) indicate that expectation is a linear operation.

Mean of a function of a random variable [Ref. 7]

The mean $E\{g(\mathbf{x})\}$ of a function $g(\mathbf{x})$ of a *discrete* random variable **x** is defined as:

$$E\{g(\mathbf{x})\} = \sum_{i=1}^{n} p_i\, g(x_i) \qquad (3.20)$$

where p_i is the probability of the event $\mathbf{x} = x_i$. If $g(\mathbf{x})$ is a function of a *continuous* random variable **x**, the mean is defined as:

$$E\{g(\mathbf{x})\} = \int_{-\infty}^{\infty} g(x) f(x)\, dx \qquad (3.21)$$

where $f(x)$ is the pdf of the random variable **x**. This result is known as the *fundamental theorem of expectation*.

Moments

The nth *moment* of the *discrete* random variable **x** is defined as:

$$m_n = E(\mathbf{x}^n) = \sum_{i=-\infty}^{\infty} x_i^n p_i \quad (3.22)$$

where p_i is the probability of the event $\mathbf{x} = x_i$. The nth *moment* of the *continuous* random variable **x** is defined as:

$$m_n = E\{\mathbf{x}^n\} = \int_{-\infty}^{\infty} x^n f(x)\, dx \quad (3.23)$$

where $f(x)$ is the pdf of **x**. Note that the moment m_1 is the same as the mean.

Central moments

The kth *central moment* μ_k of a *discrete* random variable x is defined by

$$\mu_k = \sum_{i=-\infty}^{\infty} (x_i - \bar{\mathbf{x}})^k p_i \quad (3.24)$$

where p_i is the probability of the event $\mathbf{x} = x_i$. The kth *central moment* μ_k of a *continuous* random variable **x** is defined by

$$\mu_k = E\{(\mathbf{x} - \bar{\mathbf{x}})^k\} = \int_{-\infty}^{\infty} (x - \bar{\mathbf{x}})^k f(x)\, dx \quad (3.25)$$

where $f(x)$ is the pdf of **x**.

- The second central moment is often called the *variance* and denoted by σ^2 rather than μ_n.
- The positive square root of variance is called the *standard deviation*.
- The third central moment is called the *skew*.
- The fourth central moment is called the *kurtosis*.

Properties of variance

$$\operatorname{var}\{cX\} = c^2 \operatorname{var}\{X\} \quad (3.26)$$

$$\operatorname{var}\{X\} = E\{X^2\} - (E\{X\})^2 \quad (3.27)$$

If X and Y are independent, then

$$\operatorname{var}\{X + Y\} = \operatorname{var}\{X\} + \operatorname{var}\{Y\} \quad (3.28)$$

Consider a set of random variables x_i, $i = 1, 2, \ldots, n$.

$$\text{var}\left\{\sum_{i=1}^{n} x_i\right\} = \sum_{i=1}^{n} \text{var}\{x_i\} + 2 \sum_{j=i+1}^{n} \sum_{i=1}^{n-1} \text{cov}\{x_i, x_j\} \quad (3.29)$$

Noting that

$$\text{cov}\{x_i, x_i\} = \text{var}\{x_i\} \quad (3.30)$$

and

$$\text{cov}\{x_i, x_j\} = \text{cov}\{x_j, x_i\} \quad (3.31)$$

Eq. (3.29) can be simplified to yield

$$\text{var}\left\{\sum_{i=1}^{n} x_i\right\} = \sum_{j=1}^{n} \sum_{i=1}^{n} \text{cov}\{x_i, x_j\} \quad (3.32)$$

We can relate the mean and variance of a random variable by expanding the definition of variance.

$$\begin{aligned}
\sigma^2 &= E\{(\mathbf{x} - \eta)^2\} \\
&= E\{(\mathbf{x}^2 - 2\mathbf{x}\eta + \eta^2)\} \\
&= E\{\mathbf{x}^2\} - 2\eta E\{\mathbf{x}\} + \eta^2 \\
&= E\{\mathbf{x}^2\} - 2\eta^2 + \eta^2 \\
&= E\{\mathbf{x}^2\} - [E\{\mathbf{x}\}]^2
\end{aligned}$$

Absolute and general moments

For a random variable **x**, the forms

$$E\{|x|^n\} \quad \text{and} \quad E\{|\mathbf{x} - \eta|^n\}$$

are called *absolute moments*. For a random variable **x**, the forms

$$E\{(\mathbf{x} - a)^n\} \quad \text{and} \quad E\{|\mathbf{x} - a|^n\}$$

are called *general moments*.

3.5 Characteristic Functions

The *characteristic function* $\phi_x(\omega)$ of the random variable **x** is defined as:

$$\begin{aligned}
\phi_x(\omega) &\triangleq E\{\exp(j\omega x)\} = \int_{-\infty}^{\infty} p(x) \exp(j\omega x)\, dx \\
&= \mathscr{F}^{-1}[p(x)] \quad (3.33)
\end{aligned}$$

where $p(x)$ is the pdf of **x**, and $\mathscr{F}^{-1}$ denotes the inverse Fourier transform. Some authors denote the characteristic function by $M(\omega)$ instead of $\phi(\omega)$.

If **x** and **y** are independent random variables and $\mathbf{z} = \mathbf{x} + \mathbf{y}$, then the characteristic function of **z** is given by

$$\phi_z(\omega) = \phi_x(\omega)\,\phi_y(\omega) \tag{3.34}$$

where $\phi_x(\omega)$ and $\phi_y(\omega)$ are the characteristic functions of **x** and **y**, respectively. Note that multiplication of characteristic functions corresponds to convolution of probability density functions. This follows directly from the fact that the characteristic function and pdf form a Fourier transform pair.

Relationship between moments and characteristic functions. The nth moment of a random variable **x** having the characteristic function $\phi_x(\omega)$ can be obtained from

$$m_n = \left[j^{-n} \frac{\partial^n}{\partial \omega^n} \phi_x(\omega) \right]_{\omega=0} \tag{3.35}$$

3.6 Relationships between Random Variables

If **x** and **y** are continuous random variables, their *joint distribution* $F_{xy}(x, y)$ yields the probability that **x** does not exceed x *and* **y** does not exceed y.

$$F_{xy}(x, y) = P\{\mathbf{x} \leq x, \mathbf{y} \leq y\} \tag{3.36}$$

The *joint density function*, $p_{xy}(x, y)$, of the two continuous random variables **x** and **y** is defined as the mixed partial derivative of F_{xy}.

$$p_{xy}(x, y) = \frac{\partial^2}{\partial x\, \partial y} F_{xy}(x, y) \tag{3.37}$$

The individual density functions for **x** and **y** can be obtained from $p_{xy}(x, y)$ by integration:

$$p_x(x) = \int_{-\infty}^{\infty} p_{xy}(x, y)\, dy \tag{3.38}$$

$$p_y(y) = \int_{-\infty}^{\infty} p_{xy}(x, y)\, dx \tag{3.39}$$

In this context, $p_x(x)$ and $p_y(y)$ are referred to as *marginal* probability density functions to emphasize the distinction between them and the joint density function.

If **x** and **y** are *discrete* random variables, their *joint probability function* p_{ij} is the probability that $\mathbf{x} = x_i$ and $\mathbf{y} = y_j$.

$$p_{ij} = P(\mathbf{x} = x_i, \mathbf{y} = y_j) \qquad (3.40)$$

The marginal probability functions for **x** and **y** can be obtained by summing the joint probability function over all y_j or all x_i

$$P(\mathbf{x} = x_i) = \sum_{j=-\infty}^{\infty} P(\mathbf{x} = x_i, \mathbf{y} = y_j) \qquad (3.41)$$

$$P(\mathbf{y} = y_j) = \sum_{i=-\infty}^{\infty} P(\mathbf{x} = x_i, \mathbf{y} = y_j) \qquad (3.42)$$

Statistical independence

Two random variables are called *statistically independent* if

$$P\{\mathbf{x} \in A, \mathbf{y} \in B\} = P\{\mathbf{x} \in A\}P\{\mathbf{y} \in B\} \qquad (3.43)$$

where A and B are arbitrary subsets of the ranges of **x** and **y**, respectively. This is equivalent to saying that $\{\mathbf{x} \in A\}$ and $\{\mathbf{y} \in B\}$ are independent events. If the random variables **x** and **y** are statistically independent, then the joint density $p_{xy}(x, y)$ equals the product of the marginal densities:

$$p_{xy}(x, y) = p_x(x) p_y(y) \qquad (3.44)$$

If the random variables x and y are not independent, then the joint density cannot be synthesized from just the marginal densities—either the conditional densities or sufficient other a priori information that characterizes the dependencies will be needed.

Conditional probability densities

The *conditional probability density* of **x** given that $\mathbf{y} = y$ is denoted as $p_x(x \mid \mathbf{y} = y)$. When the meaning is clear, this is often abbreviated as $p(x \mid y)$. The conditional pdf $p(x \mid y)$ and joint pdf $p(x, y)$ are related by

$$P(x \mid y) = \frac{p(x, y)}{p(y)} \qquad (3.45)$$

If **x** and **y** are statistically independent, then

$$p(x, y) = p(x) p(y) \qquad (3.46)$$

$$p(x \mid y) = p(x) \qquad (3.47)$$

Joint moments

The *joint moment* m_{kn} of the two *continuous* random variables **x** and **y** is defined by

$$m_{kn} = E\{\mathbf{x}^k \mathbf{y}^n\} = \int_{-\infty}^{\infty} \int_{-\infty}^{\infty} x^k y^n p_{xy}(x, y)\, dx\, dy \qquad (3.48)$$

The moment m_{kn} is of *order* $k + n$. The *joint central moment* μ_{kn} of the two random variables **x** and **y** is defined by

$$\mu_{kn} = E\{(\mathbf{x} - \bar{\mathbf{x}})^k (\mathbf{y} - \bar{\mathbf{y}})^n\} \qquad (3.49)$$

The *joint moment* m_{kn} of the two *discrete* random variables **x** and **y** is defined by

$$m_{kn} = E\{\mathbf{x}^k \mathbf{y}^k\} = \sum_{i=-\infty}^{\infty} \sum_{j=-\infty}^{\infty} x_i^k y_j^n p_{ij} \qquad (3.50)$$

3.7 Correlation and Covariance

The *correlation* R_{xy} of two random variables **x** and **y** is defined as:

$$R_{xy} = E[\mathbf{xy}] = \int_{-\infty}^{\infty} \int_{-\infty}^{\infty} xy\, p_{xy}(x, y)\, dx\, dy \qquad (3.51)$$

The *covariance* C_{xy} of two random variables **x** and **y** is defined as:

$$C_{xy} = E\{(\mathbf{x} - \eta_x)(\mathbf{y} - \eta_y)\} \qquad (3.52)$$

It can be shown that $|C_{xy}| \leq \sigma_x \sigma_y$. In some applications, it is more convenient to use a normalized measure called the *correlation coefficient*, which is defined as:

$$r_{xy} = \frac{C_{xy}}{\sigma_x \sigma_y} \qquad (3.53)$$

where C_{xy} = covariance of **x** and **y**
σ_x = standard deviation of **x**
σ_y = standard deviation of **y**

It can be shown that $|r_{xy}| \leq 1$.

Two random variables are called *uncorrelated* if their covariance and correlation coefficients are both zero. (Either one being zero is a sufficient condition for uncorrelatedness and implies that the other is also zero.) If **x** and **y** are uncorrelated, then

$$E\{\mathbf{xy}\} = E\{\mathbf{x}\}E\{\mathbf{y}\} \qquad (3.54)$$

and

$$\sigma_{x+y}^2 = \sigma_x^2 + \sigma_y^2 \qquad (3.55)$$

If two random variables **x** and **y** are statistically independent, then they are also uncorrelated. However, two RVs may be uncorrelated but not independent. In the case of normal random variables, uncorrelatedness *is* sufficient to establish statistical independence.

Two random variables **x** and **y** are *orthogonal* if and only if $E\{\mathbf{xy}\} = 0$. The orthogonality between **x** and **y** can be denoted as $\mathbf{x} \perp \mathbf{y}$. If **x** and **y** are *uncorrelated*, then

$$(\mathbf{x} - \bar{\mathbf{x}}) \perp (\mathbf{y} - \bar{\mathbf{y}}) \qquad (3.56)$$

If **x** and **y** are uncorrelated and either or both has zero mean, then **x** is orthogonal to **y**.

The various covariance relationships between random variables can be summarized as follows:

$\operatorname{cov}(x, y) = 0$ implies that x and y are uncorrelated.

$E(x, y) = E(x)E(y)$ is equivalent to x and y being uncorrelated.

x and y statistically independent implies that x and y are uncorrelated.

$p_{xy}(x, y) = p_x(x)p_y(y)$ is equivalent to x and y being statistically independent.

3.8 Probability Densities for Functions of a Random Variable

Given:

X is a continuous RV with pdf $p_x(X)$.

Y is a monotonically increasing or monotonically decreasing function of X that is differentiable for all values of X.

Then the pdf of Y is given by

$$p_y(Y) = p_x(Y[X]) \cdot \left| \frac{dY}{dX} \right| \qquad (3.57)$$

where $Y[X] \triangleq f^{-1}(y)$ given that $y = f(x)$

Example Let X be a random variable uniformly distributed between 0 and 1:

$$p_x(X) = \begin{cases} 1 & 0 \le X \le 1 \\ 0 & \text{otherwise} \end{cases}$$

Find the pdf for $y = -\ln x$.

solution It follows directly that

$$Y[X] = \exp(-Y) \quad (3.58)$$

$$\left|\frac{dY}{dX}\right| = -\exp(-Y) \quad (3.59)$$

Substituting (3.58) and (3.59) into (3.57) yields

$$p_y(Y) = p_x[\exp(-Y)] \cdot |-\exp(-Y)| \quad (3.60)$$

We note that $0 \le \exp(-Y) \le 1$ for all $Y \ge 0$; thus

$$p_x[\exp(-Y)] = \begin{cases} 1 & Y \ge 0 \\ 0 & Y < 0 \end{cases}$$

Therefore (3.60) simplifies to

$$p_y(Y) = \begin{cases} \exp(-Y) & Y \ge 0 \\ 0 & Y < 0 \end{cases}$$

3.9 Probability Densities for Functions of Two Random Variables

If X and Y are random variables, and U and V are random variables obtained from X and Y by

$$U = g_1(X, Y) \quad (3.61)$$

$$V = g_2(X, Y) \quad (3.62)$$

then the joint pdf of U and V is obtained from the joint pdf of X and Y using

$$p_{uv}(U, V) = \frac{p_{xy}(X, Y)}{|J|}\bigg|_{(X, Y) = g^{-1}(U, V)} \quad (3.63)$$

where J is given by

$$J = \det \begin{bmatrix} \frac{\partial}{\partial X}g_1 & \frac{\partial}{\partial Y}g_1 \\ \frac{\partial}{\partial X}g_2 & \frac{\partial}{\partial Y}g_2 \end{bmatrix} \quad (3.64)$$

The strange notation in (3.63) means, "Solve (3.61) and (3.62) for X and Y, then plug the results into $p_{xy}(x, y)$."

Probability density function for a linear combination of two random variables

Define a random variable Z as a linear combination of two other random variables X and Y:

$$Z = aX + bY$$

where a and b are real constants. The probability density function for Z can be obtained from the joint pdf for X and Y:

$$p_z(Z) = \frac{1}{|a|} \int_{-\infty}^{\infty} p_{xy}\left(\frac{Z - bY}{a}, Y\right) dY \qquad (3.65)$$

Probability distribution for a product of discrete random variables

Consider two discrete random variables X and Y. Define a new random variable Z which equals the product XY. The probability function for Z is given by

$$P(Z = z_k) = \sum_{\forall (i,j) \,|\, x_i y_j = z_k} P(X = x_i, Y = y_j) \qquad (3.66)$$

The summation in (3.66) is performed over all combinations of (i, j) for which the product $x_i y_j$ equals z_k. The summand is the joint probability that x equals i and y equals j. An alternative (and perhaps clearer) way to write (3.66) is given by

$$P(Z = z_k) = \sum_{\text{all } (i,j)} Q_{ij} \qquad (3.67)$$

where $Q_{ij} = \begin{cases} P(X = x_i, Y = y_j) & \text{for } x_i y_j = z_k \\ 0 & \text{otherwise} \end{cases}$

3.10 References

1. I. N. Gibra: *Probability and Statistical Inference for Scientists and Engineers*, Prentice-Hall, Englewood Cliffs, N.J., 1973
2. A. Papoulis: *Probability, Random Variables, and Stochastic Processes*, 2d ed., McGraw-Hill, New York, 1984.
3. A. D. Whalen: *Detection of Signals in Noise*, Academic Press, New York, 1971.
4. H. Urkowitz: *Signal Theory and Random Processes*, Artech House, Dedham, Mass., 1983.
5. D. E. Knuth: *The Art of Computer Programming: Vol. 2/Seminumerical Algorithms*, 2d ed., Addison-Wesley, Reading, Mass., 1981.
6. J. G. Proakis: *Digital Communications*, McGraw-Hill, New York, 1983.
7. W. A. Gardner: *Introduction to Random Processes*, Macmillan, New York, 1986.

Chapter 4

Probability Distributions in Communications

4.1 Gaussian Random Variables

A zero-mean, unity variance *gaussian* random variable y has a probability density function given by

$$p(y) = \frac{1}{\sqrt{2\pi}} \exp \frac{-y^2}{2} \qquad (4.1)$$

A sketch of (4.1) is shown in Fig. 4.1. The corresponding cumulative distribution function is obtained by integrating (4.1).

$$P(y \leq Y) = \frac{1}{\sqrt{2\pi}} \int_{-\infty}^{Y} e^{-y^2/2} \, dy \qquad (4.2)$$

A sketch of (4.2) is shown in Fig. 4.2. The integral in (4.2) cannot be evaluated in closed form; but it occurs so often in engineering and science that a special function, called the *error function* and denoted as erf, has been defined as:

$$\text{erf } x = \frac{1}{\sqrt{2\pi}} \int_{0}^{x} e^{-y^2/2} \, dy \qquad (4.3)$$

Closed-form approximations for the error function have been established (see Sec. 2.8), and values for erf x along with other closely related functions have been extensively tabulated. Thus $P(y \leq Y)$ can be obtained as:

$$P(y \leq Y) = 0.5 + \text{erf } y \qquad (4.4)$$

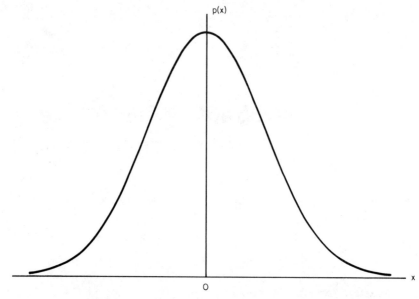

Figure 4.1 Gaussian probability density function.

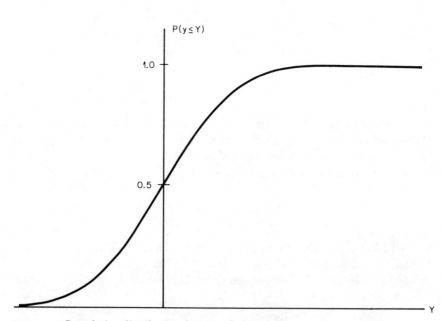

Figure 4.2 Cumulative distribution function for a gaussian random variable.

For a gaussian random variable of mean μ and variance σ^2, the pdf given in (4.1) can be scaled by σ and shifted by μ to yield

$$p(y) = \frac{1}{\sigma\sqrt{2\pi}} e^{-(y-\mu)^2/2\sigma^2} \qquad (4.5)$$

The characteristic function $\phi_y(\omega)$ of the gaussian random variable y is given by

$$\phi_y(\omega) = \exp\frac{j\mu\omega - \sigma^2\omega^2}{2} \qquad (4.6)$$

The gaussian distribution is named in honor of Johann Karl Friedrich Gauss (1777–1855), a German mathematician who is widely regarded as perhaps the greatest mathematician of all time. A gaussian random variable is also called a *normal variate*. Engineering literature seems to favor the use of "gaussian," while the mathematical literature favors "normal"—perhaps this is due to the multitude of other things to which mathematicians attach the name of Gauss. A normal variate of mean μ and variance σ^2 is sometimes denoted as $N(\mu; \sigma^2)$.

Jointly gaussian random variables. Consider a set of random variables, $x_1, x_2, \ldots, x_n$, which have means $\mu_1, \mu_2, \ldots, \mu_n$, respectively. These random variables are described as *jointly gaussian* if their joint probability density function can be given by

$$p(\mathbf{x}) = \frac{1}{\sqrt{(2\pi)^n |\mathbf{k}|}} \exp\left[\frac{(\mathbf{x}-\boldsymbol{\mu})^T \mathbf{k}^{-1}(\mathbf{x}-\boldsymbol{\mu})}{-2}\right] \qquad (4.7)$$

where $\mathbf{x} = \begin{bmatrix} x_1 \\ x_2 \\ \vdots \\ x_n \end{bmatrix} \qquad \boldsymbol{\mu} = \begin{bmatrix} \mu_1 \\ \mu_2 \\ \vdots \\ \mu_n \end{bmatrix}$

$$\mathbf{k} = \begin{bmatrix} \text{cov}(x_1, x_1) & \text{cov}(x_1, x_2) & \cdots & \text{cov}(x_1, x_n) \\ \text{cov}(x_2, x_1) & \text{cov}(x_1, x_2) & \cdots & \text{cov}(x_2, x_n) \\ \vdots & \vdots & \ddots & \vdots \\ \text{cov}(x_n, x_1) & \text{cov}(x_n, x_2) & \cdots & \text{cov}(x_n, x_n) \end{bmatrix}$$

$|\mathbf{k}|$ denotes the determinant of $\mathbf{k}$

Example Find the joint pdf for two zero-mean, jointly gaussian random variables x_1 and x_2.

solution

$$\mathbf{x} = \begin{bmatrix} x_1 \\ x_2 \end{bmatrix} \quad \boldsymbol{\mu} = \begin{bmatrix} 0 \\ 0 \end{bmatrix} \quad \mathbf{k} = \begin{bmatrix} \sigma_1^2 & C_{12} \\ C_{21} & \sigma_2^2 \end{bmatrix}$$

$$|\mathbf{k}| = \det \mathbf{k} = \sigma_1^2 \sigma_2^2 - C_{12} C_{21} = \sigma_1^2 \sigma_2^2 - C_{12}^2$$

where $\sigma_1^2 = \text{var}(x_1)$
$\sigma_2^2 = \text{var}(x_2)$
$C_{12} = \text{cov}(x_1, x_2) = \text{cov}(x_2, x_1) = C_{21}$

Using any convenient matrix inversion technique, $\mathbf{k}^{-1}$ can be found to be

$$\mathbf{k}^{-1} = \begin{bmatrix} \dfrac{\sigma_2^2}{\sigma_1^2 \sigma_2^2 - C_{12} C_{21}} & \dfrac{-C_{12}}{\sigma_1^2 \sigma_2^2 - C_{12} C_{21}} \\ \dfrac{-C_{21}}{\sigma_1^2 \sigma_2^2 - C_{12} C_{21}} & \dfrac{\sigma_1^2}{\sigma_1^2 \sigma_2^2 - C_{12} C_{21}} \end{bmatrix}$$

Making the appropriate substitutions, we find

$$p(x_1 x_2) = \frac{1}{2\pi \sqrt{\sigma_1^2 \sigma_2^2 - C_{12}^2}} \exp\left[\frac{x_1^2 \sigma_2^2 - 2 x_1 x_2 C_{12} + x_2^2 \sigma_1^2}{-2(\sigma_1^2 \sigma_2^2 - C_{12}^2)}\right]$$

4.2 Uniform Random Variables

A uniformly distributed random variable x has a probability density function given by

$$p(x) = \begin{cases} \dfrac{1}{b - a} & a \leq x \leq b \\ 0 & \text{elsewhere} \end{cases} \tag{4.8}$$

The mean is given by

$$\mu = \frac{a + b}{2} \tag{4.9}$$

and the variance is given by

$$\sigma^2 = \frac{(b - a)^2}{12} \tag{4.10}$$

The characteristic function $\phi_x(\omega)$ of a uniform random variable x with pdf as in (4.8) is given by

$$\phi_x(\omega) = \frac{\exp(j\omega b) - \exp(j\omega a)}{j\omega(b - a)} \tag{4.11}$$

4.3 Exponential Random Variables [Refs. 1 through 3]

An exponentially distributed random variable has a pdf given by

$$p(x) = \begin{cases} \lambda \exp(-\lambda x) & x \geq 0 \\ 0 & x < 0 \end{cases} \quad (4.12)$$

The mean and variance are given by

$$\mu_x = \lambda^{-1} \quad (4.13)$$

$$\sigma_x^2 = \lambda^{-2} \quad (4.14)$$

Note that the mean and variance cannot be independently specified, as they are both completely determined by the single parameter λ. The cumulative distribution function for an exponential random variable is given by

$$P(x \leq X) = 1 - \exp(-\lambda X) \quad (4.15)$$

The characteristic function of an exponentially distributed random variable is given by

$$\phi_x(\omega) = \frac{\lambda}{\lambda - 2j\omega} \quad (4.16)$$

Relationship between uniform and exponential random variables. An exponential random variable can be obtained as the negative natural logarithm of a random variable uniformly distributed on [0, 1]. Consider a uniform random variable U having a pdf given by

$$p(U) = \begin{cases} 1 & 0 \leq U \leq 1 \\ 0 & \text{otherwise} \end{cases}$$

Define a new random variable $X = -\ln U$. Using Eq. (3.57), the pdf of X is found to be

$$p(X) = \begin{cases} \exp(-X) & 0 \leq U \leq 1 \\ 0 & \text{otherwise} \end{cases}$$

4.4 Rayleigh Random Variables [Refs. 3 and 4]

A Rayleigh random variable has a probability density function given by

$$p(x) = \begin{cases} \dfrac{x}{b^2} \exp\left(\dfrac{-x^2}{2b^2}\right) & x \geq 0 \\ 0 & x < 0 \end{cases} \quad (4.17)$$

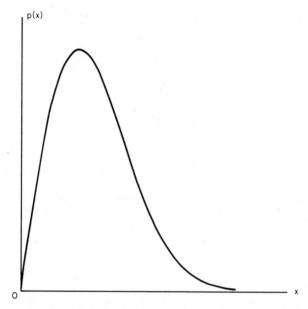

Figure 4.3 Probability density function for a Rayleigh random variable.

A sketch of (4.17) is shown in Fig. 4.3. The mean and variance of a Rayleigh random variable are given by

$$\mu = b\sqrt{\frac{\pi}{2}} \qquad (4.18)$$

$$\sigma^2 = \frac{(4-\pi)b^2}{2} \qquad (4.19)$$

Note that the mean and variance cannot be independently specified, as they are both completely determined by the single parameter b. The cumulative distribution function of a Rayleigh random variable is given by

$$P(x \leq X) = 1 - \exp\left(\frac{-X^2}{2b^2}\right) \qquad (4.20)$$

A sketch of (4.20) is shown in Fig. 4.4. The characteristic function of a Rayleigh random variable is given by

$$\phi_x(\omega) = 1 + jb\omega\sqrt{\frac{\pi}{2}}\left[1 + \text{erf}\left(\frac{jb\omega}{\sqrt{2}}\right)\right]\exp\left(\frac{-b^2\omega^2}{2}\right) \qquad (4.21)$$

The Rayleigh distribution is named for John William Strutt (1842–1919), an English physicist who was the third Baron Rayleigh.

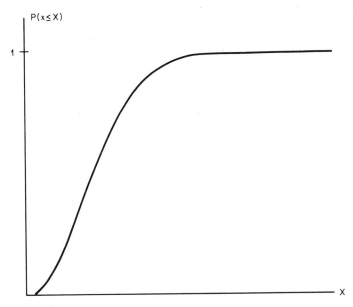

Figure 4.4 Cumulative distribution function for a Rayleigh random variable.

Relationship between Rayleigh and exponential random variables. The square of a Rayleigh random variable is an exponential random variable. Consider a Rayleigh random variable R having a pdf, mean, and variance given by

$$p_R(R) = \frac{R}{b^2} \exp\left(\frac{-R^2}{2b^2}\right) \quad \text{for } R \geq 0 \qquad (4.22)$$

$$\mu_R = b\sqrt{\frac{\pi}{2}} \qquad (4.23)$$

$$\sigma_R^2 = \frac{(4-\pi)b^2}{2} \qquad (4.24)$$

Define a new random variable $X = R^2$. Using Eq. (3.57), the pdf of X is found to be

$$p_x(X) = \frac{1}{2b^2} \exp\left(\frac{-X^2}{2b^2}\right) \quad \text{for } X \geq 0 \qquad (4.25)$$

The mean and variance of X are given by

$$\mu_x = 2b^2 \qquad (4.26)$$

$$\sigma_x^2 = 4b^4 \qquad (4.27)$$

4.5 Rice Random Variables [Refs. 3 and 4]

A Rice random variable has a pdf given by

$$p(x) = \begin{cases} \dfrac{x}{b^2} \exp\left(\dfrac{x^2+a^2}{-2b^2}\right) I_0\left(\dfrac{ax}{b^2}\right) & x \geq 0 \\ 0 & \text{otherwise} \end{cases} \qquad (4.28)$$

where $I_0(\cdot)$ denotes the modified zero-order Bessel function of the first kind (see Sec. 2.8). A sketch of (4.28) is shown in Fig. 4.5. The mean of a Rice random variable is given by

$$\mu = \frac{b\sqrt{2\pi}}{2} {}_1F_1\left[\frac{-1}{2}; 1; \frac{-(a/b)^2}{2}\right] \qquad (4.29)$$

where ${}_1F_1(\cdot\,;\,\cdot\,;\,\cdot)$ denotes the confluent hypergeometric function. A plot of μ/b versus a/b is shown in Fig. 4.6. The variance of a Rice random variable is given by

$$\sigma^2 = 2b^2 {}_1F_1\left[-1; 1; \frac{(a/b)^2}{2}\right] - \mu^2 \qquad (4.30)$$

A plot of σ/b versus a/b is shown in Fig. 4.7. Note that for $a = 0$, the pdf in (4.28) equals the pdf of a Rayleigh random variable. Sketches

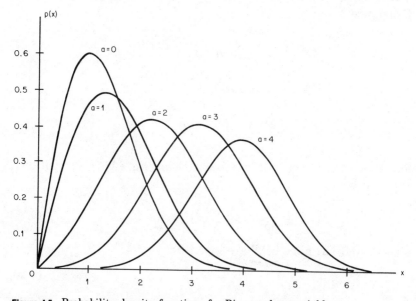

Figure 4.5 Probability density functions for Rice random variables.

Probability Distributions in Communications   53

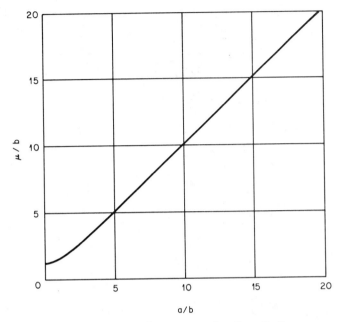

Figure 4.6 Rice distribution's mean as a function of a/b.

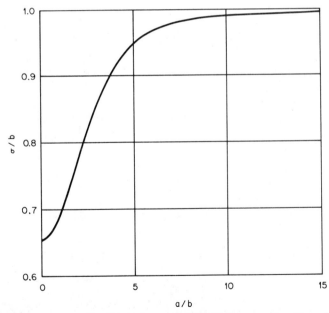

Figure 4.7 Rice distribution's standard deviation as a function of a/b.

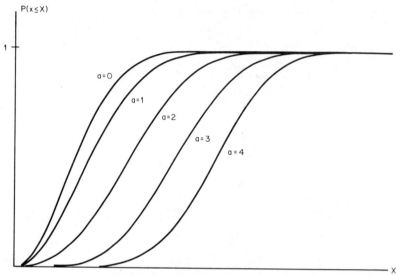

Figure 4.8 Cumulative distribution functions for Rice random variables.

of the cumulative distribution functions for various values of a are shown in Fig. 4.8. The Rice distribution is sometimes called a *generalized Rayleigh* distribution.

4.6 Chi-Squared Random Variables [Refs. 1, 3, and 4]

The probability density function for a chi-squared distribution with n degrees of freedom is given by

$$p(x) = \begin{cases} \dfrac{x^{n/2-1} \exp(-x/2)}{\sqrt{2^n}\,\Gamma(n/2)} & x \geq 0 \\ 0 & x < 0 \end{cases} \qquad (4.31)$$

The mean and variance are given by

$$\mu = n \qquad (4.32)$$

$$\sigma^2 = 2n \qquad (4.33)$$

Note that the mean and variance cannot be independently specified, as they are both completely determined by the single parameter n.

4.7 Poisson Distribution

A Poisson distribution is a discrete-type distribution with a probability function given by

$$P\{x = k\} = \exp(-a)\frac{a^k}{k!} \qquad k = 0, 1, 2, \ldots$$

The mean and variance of such a distribution are each equal to the parameter a

$$\mu = a \qquad \sigma^2 = a$$

4.8 Gamma and Erlang Distributions

The gamma distribution has a pdf given by

$$p(x) = \frac{c^{b+1}}{\Gamma(b+1)} x^b \exp(-cx) \qquad x \geq 0, c > 0$$

where $\Gamma(b) \equiv \int_0^\infty y^{b-1} e^{-y} dy$

An Erlang distribution is simply a gamma distribution in which b is an integer. The gamma function then reduces to a factorial, and the pdf becomes

$$p(x) = \frac{c^n}{(n-1)!} x^{n-1} \exp(-cx) \qquad x \geq 0$$

Note that for $n = 1$, the Erlang distribution reduces to an exponential distribution. The mean and variance of the Erlang distribution are given by

$$\mu = \frac{n}{c} \qquad \sigma^2 = \frac{n}{c^2}$$

4.9 REFERENCES

1. J. N. Gibra: *Probability and Statistical Inference for Scientists and Engineers*, Prentice-Hall, Englewood Cliffs, N.J., 1973.
2. A. Papoulis: *Probability, Random Variables, and Stochastic Processes*, 2d ed., McGraw-Hill, New York, 1984.
3. A. D. Whalen: *Detection of Signals in Noise*, Academic Press, New York, 1971.
4. H. Urkowitz, *Signal Theory and Random Processes*, Artech House, Dedham, Mass., 1983.

Chapter 5

Random Processes

5.1 Random Processes [Refs. 2 and 3]

Similar to the way in which a random variable assigns *numeric values* to experimental outcomes, a *random process* assigns *functions* to experimental outcomes. The functions are called *sample functions*, *member functions*, or *realizations* of the random process. In most of the engineering literature, *sample function* predominates; however, in works (such as this one) that deal with signal processing, it is probably better to use "realization" so as to avoid confusion between "sample function" and either "sampling function" or "sampled function." In the majority of engineering applications, each realization is a function of time (usually) or space (sometimes). Consider the collection, or *ensemble*, of sample functions shown in Fig. 5.1.

Each realization $\mathbf{x}(t, \alpha_n)$ has a time average given by

$$\langle \mathbf{x}(t, \alpha_n) \rangle = \lim_{T \to \infty} \frac{1}{2T} \int_{-T}^{T} \mathbf{x}(t, \alpha_n) \, dt$$

This average is a random variable that takes on different values as different realizations $\mathbf{x}(t, \alpha_1)$, $\mathbf{x}(t, \alpha_2)$, and so on are substituted for $\mathbf{x}(t, \alpha_n)$.

Conceptually, at any instant of time, we could measure the instantaneous values of each realization in the ensemble. The set of values so obtained forms a random variable which can be statistically characterized apart from the characterization of the process itself. For clarity we will sometimes refer to such a random variable as an *ensemble random variable* even though such terminology is not widespread. Thus a random process can be viewed as a time-indexed family of random variables, which in the case of a discrete-time

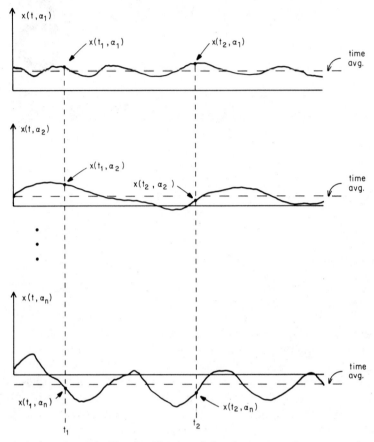

Figure 5.1 An ensemble of realizations belonging to a random process.

process, becomes a time sequence of random variables. In general, the statistics of the ensemble random variables are unrelated to the time statistics of the realizations. However, in *ergodic* random processes, the ensemble statistics and time statistics are interchangeable.

If the realizations are themselves random in nature (such as shown above), the process is called a *regular random process*. A regular random process could be used to model the position of sand grains in a sandstorm. The position of each grain is described by a realization of the random process. Since the future position of a grain cannot be expressed exactly in terms of its past position, each sample function is random. Likewise, knowing the position of a particular grain at a specific instant in time will not give us the position of any other grains. Thus the ensemble of positions at a fixed instant in time forms a random variable.

If the sample functions are deterministic, the process is called a *predictable random process*. An example of a predictable random process would be a group of RC oscillators assembled from randomly selected components. The amplitude, phase, and frequency of the oscillators will be random variables mapped from the specific characteristics of the selected components. However, unlike the motion of a sand particle, the future output of a particular oscillator can be determined from its past output.

It should be noted that usually an ensemble is not a collection of actual sample functions; rather, it is an abstract concept which proves useful in the statistical characterization of random phenomena. Instead of having a large collection of oscillators, we may actually have only one oscillator. Prior to switching it on, we can still characterize its behavior based on the statistics of a hypothetical collection of oscillators which could be built from the available set of components. However, once it is turned on and observed for a while, it may be more appropriate to discard the statistical characterization and instead describe the oscillator output as a deterministic function of time.

5.2 Stationarity

As discussed in Sec. 5.1, the instantaneous values of all the realizations comprising a random process will, when taken at any single instant of time, constitute a random variable. This *ensemble random variable* (ERV) will have statistics which are in general unrelated to the statistics over time of the individual realizations.

Furthermore, the statistics of the ERV at one instant may be quite different from the statistics of the ERV at a different instant. In fact, at each instant of time an ERV with potentially different statistics can be formed. Joint statistics will of course be exhibited within any group of two or more ERVs.

Many applications of practical interest will involve random processes in which the ERVs at each instant individually exhibit identical statistics. The term *stationary* is used to describe a random process in which individual ERV statistics as well as all possible joint statistics are time invariant, that is,

$$p(x_1, x_2, \ldots, x_n; t_1, t_2, \ldots, t_n)$$
$$= p(x_{1+\tau}, x_{2+\tau}, \ldots, x_{n+\tau}; t_{1+\tau}, t_{2+\tau}, \ldots, t_{n+\tau}) \quad \text{for all } \tau$$

Knowledge that a particular process is stationary can be very useful; but unfortunately, proving the time invariance of all possible joint statistics is almost always difficult and usually impossible.

Wide-sense stationarity is a weaker form of invariance which is easier to establish than strict stationarity, yet almost as useful. A random process is wide-sense stationary (often abbreviated as wss) if its mean is constant over time and the autocorrelation R satisfies

$$R(t + \tau, t) = R(\tau, 0) \quad \text{for all } t \text{ and } \tau$$

Sometimes (especially in British literature) a wss process is called *weakly stationary*. A random process is called *covariance stationary* if its autocovariance function K (see Sec. 5.3) satisfies

$$K(t + \tau, t) = K(\tau, 0) \quad \text{for all } t \text{ and } \tau$$

Wide-sense stationarity implies covariance stationarity.

5.3 Autocorrelation and Autocovariance

The *autocorrelation function* (acf) of a random process is denoted by $R_x(t_1, t_2)$ and is obtained by forming the cross-correlation of the ensemble random variables at time t_1 and t_2.

$$R_x(t_1, t_2) = E[x_1 x_2] = \int_{-\infty}^{\infty} \int_{-\infty}^{\infty} x_1 x_2^* p_{x_1 x_2}(x_1, x_2) \, dx_1 \, dx_2 \quad (5.1)$$

where x_1 represents $x(t_1)$ and x_2 represents $x(t_2)$. In general, $x(t)$ can be complex-valued, and the superscript asterisk denotes complex conjugation. Most texts assume that $x(t)$ is real-valued and define the autocorrelation without the conjugation operator shown here. The acf of a discrete random process is defined by

$$R_x(t_1, t_2) = \sum_{i=1}^{N} \sum_{j=1}^{N} x(t_1) x^*(t_2) P[x(t_1) = a_i, x(t_2) = a_j] \quad (5.2)$$

If $R_x(t_1, t_2)$ is a function of only the difference $t_2 - t_1$ and not of the specific values of t_1 and t_2, the RV x is wide-sense stationary and the autocorrelation can be denoted as:

$$R_x(\tau) = E[\mathbf{x}(t)\mathbf{x}(t + \tau)] \quad (5.3)$$

At least one well-known text [Ref. 4] uses the term *autocorrelation function* to describe the normalized *autocovariance function* given by

$$p_x(t_1, t_2) = \frac{R_x(t_1, t_2)}{\sigma_x(t_1)\sigma_x(t_2)} \quad (5.4)$$

Note that (5.4) is analogous to the correlation coefficient defined for random variables in Eq. (3.38).

Properties of autocorrelation functions
1. The autocorrelation function is nonnegative.
2. The power in the process x is given by

$$\text{Power} = E\{x^2(t)\} = R_x(t, t) = \sigma_x^2$$

3. A second-order process is one for which $R_x(t, t) < \infty$ for all t.
4. $R_x(t, s) = R_x(s, t)$ for all t and s.
5. $\quad\quad\quad\quad |R_x(t, s)| \leq \sqrt{R_x(t, t) R_x(s, s)}$
6. $R_x(t, s)$ is nonnegative definite. In other words, for all n and all $t_1, t_2, t_3, \ldots, t_n$, there are complex numbers $\alpha_1, \alpha_2, \alpha_3, \ldots, \alpha_n$ such that

$$\sum_{j=1}^{n} \sum_{k=1}^{m} \alpha_j \alpha_k^* R_x(t_j, t_k) \geq 0 \quad \text{for } j, k \in \{t_1, t_2, \ldots, t_n\}$$

7. For wss processes, $R_x(\tau) = R_x(-\tau)$.
8. For wss processes, $|R_x(\tau)| \leq R_x(0)$.

Autocovariance

The autocovariance function (acvf) of a random process is denoted by $K_x(t_1, t_2)$ and is obtained by forming the covariance of the ensemble random variables at time t_1 and t_2.

$$K_x(t_1, t_2) = E\{[\mathbf{x}(t_1) - m_x(t_1)][\mathbf{x}(t_2) - m_x(t_2)]\} \quad (5.5)$$

Expansion of (5.5) yields an expression which relates autocovariance, autocorrelation, and mean:

$$K_x(t_1, t_2) = R_x(t_1, t_2) - m_x(t_1) m_x(t_2) \quad (5.6)$$

Uncorrelated random processes

A pair of random processes $x(t)$ and $y(t)$ is described as *uncorrelated* if for all possible t and s:

$$E[x(t)y(s)] = E[x(t)]E[y(s)] \quad (5.7)$$

or equivalently if

$$R_{xy}(t, s) = \mu_x(t) \mu_y(s) \quad (5.8)$$

A single random process is described as *uncorrelated* if its ensemble random variables are uncorrelated with each other (see Sec. 3.7). A random process is described as *independent* if its constituent ensemble random variables are independent.

5.4 Power Spectral Density of Random Processes

The *power spectral density* $S_x(f)$ and the *autocorrelation function* $R_x(\tau)$ of a wide-sense stationary random process $X(t)$ comprise a Fourier transform pair.

$$S_x(f) = \mathscr{F}\{R_x(\tau)\} = \int_{-\infty}^{\infty} R_x(\tau)\exp(-j2\pi ft)\,d\tau \qquad (5.9)$$

$$R_x(\tau) = \mathscr{F}^{-1}\{S_x(f)\} = \int_{-\infty}^{\infty} S_x(f)\exp(j2\pi ft)\,df \qquad (5.10)$$

In some of the literature, the autocorrelation and power spectral density (psd) are developed separately, and then the fact that they form a Fourier transform pair is demonstrated and dubbed the *Wiener-Khintchine theorem*. Some texts take the alternative approach of developing the autocorrelation function and then *defining* the psd as the Fourier transform of the acf. In this case, (5.9) and (5.10) are called the *Wiener-Khintchine relations* since they are defined rather than derived.

There is some disagreement concerning the proper spelling of "Khintchine." The spelling used here agrees with Refs. 5 and 7. Other observed spellings include "Kinchin" [Ref. 2], "Khinchin" [Ref. 8], "Kinchine" [Refs. 9 and 10], and "Khinchine" [Ref. 6]. This is somewhat understandable due to various transliterations from Cyrillic to Latin alphabets, but there is even some disagreement over "Wiener" which in at least two texts [Refs. 6 and 7] appears as "Weiner." At least one text [Ref. 1] avoids the issue completely by presenting the relationship but not giving it a name.

5.5 Linear Filtering of Random Processes
[Refs. 1, 5, and 6]

Application of a random process $x(t)$ to the input of a linear time-invariant filter will produce a random process $Y(f)$ at the filter output. Some statistical properties of the output can be determined directly from the filter's impulse response and the statistical properties of the input. For an input process having a mean of $\mu_x(t)$, the output will have a mean given by

$$\mu_y(t) = \int_{-\infty}^{\infty} h(\tau)\mu_x(t-\tau)\,d\tau \qquad (5.11)$$

For the special case of $x(t)$ being wide-sense stationary, the mean of the output can be simplified to $\mu_y = \mu_x H(0)$. For the general case of an input process with an autocorrelation of $R_x(t_1, t_2)$, the autocorrelation of the output will be given by

$$R_y(t_1, t_2) = \int_{-\infty}^{\infty} h(\tau_1)\,d\tau_1 \int_{-\infty}^{\infty} h(\tau_2) R_x(t_1-\tau_1, t_2-\tau_2)\,d\tau_2 \qquad (5.12)$$

For the special case of $x(t)$ being wide-sense stationary, the autocorrelation can be simplified to

$$R_y(\tau) = \int_{-\infty}^{\infty}\int_{-\infty}^{\infty} h(\tau_1) h(\tau_2) R_x(t - \tau_1 + \tau_2)\,d\tau_1\,d\tau_2 \qquad (5.13)$$

For an input with power spectral density of $S_x(f)$, the output will have a power spectral density given by

$$S_y(f) = |H(f)|^2 S_x(f) \qquad (5.14)$$

5.6 Gaussian Random Processes

Gaussian random processes exhibit the following properties:

1. When any linear operation is performed upon a gaussian random process, the resulting random process is also gaussian.
2. All wide-sense stationary gaussian random processes are also strict-sense stationary.

Conditional pdf of two sample functions from a gaussian random process

Assume that $x(t)$ and $x(t + \tau)$ are sample functions from a normally distributed, zero-mean, real-valued random process. The conditional pdf of $x(t + \tau)$ given $x(t)$ is given by

$$p[x(t+\tau)\,|\,x(t)] = \left\{2\pi R(0)\left[1 - \left(\frac{R(\tau)}{R(0)}\right)^2\right]\right\}^{-1/2}$$

$$\cdot \exp\left[\frac{\{x(t+\tau) - x(t)R(\tau)/R(0)\}^2}{2R(0)\{1 - [R(\tau)/R(0)]^2\}}\right]$$

where $R(\tau)$ is the autocorrelation of the random process.

Joint pdf of two sample functions from a gaussian random process

Assume that $x(t)$ and $x(t + \tau)$ are sample functions from a normally distributed, zero-mean, real random process. The joint pdf of $x(t)$ and $x(t + \tau)$ is given by

$$p[x(t + \tau), x(t)] = \frac{1}{2\pi R(0)\sqrt{1 - [R(\tau)/R(0)]^2}}$$
$$\cdot \exp\left\{\frac{-[x(t)]^2 + 2[R(\tau)/R(0)]x(t)x(t + \tau) - [x(t + \tau)]^2}{2R(0)[1 - [R(\tau)/R(0)]^2]}\right\}$$

where $R(\tau)$ is the autocorrelation function of the random process.

5.7 Markov Processes

Within the literature there appears to be some confusion and disagreement concerning the terminology used to describe *Markov processes* and *Markov chains*. The terminology used by Papoulis [Ref. 3] appears to be the most detailed and explicitly descriptive of any of the common variations. Except where otherwise noted, the terminology which will be presented and used within this book will be similar to the terminology of Papoulis.

A *Markov process* can be described as a stochastic process whose future depends (in a probabilistic sense) only upon its present—for a given present condition of the process, the past behavior of the process will have no impact upon its future behavior.

Markov processes can be divided into four categories: (1) continuous-time, continuous-valued; (2) discrete-time, continuous-valued; (3) continuous-time, discrete-valued; and (4) discrete-time, discrete-valued. Discrete-valued Markov processes are usually referred to as *Markov chains*. Kemeny and Snell [Ref. 16] refer to discrete-valued Markov processes as *finite Markov processes*; they reserve the term "Markov chain" for something we will discuss shortly which is more properly called a *homogeneous Markov process*.

In this book we will be concerned primarily with Markov chains. General Markov processes are mentioned only as a matter of general interest since they are the parent structure within which Markov chains form a specific subset. Markov processes are named after the Russian mathematician A. A. Markov (1856–1922). (Papoulis and some others prefer to spell it Markoff.)

Markov chains

Markov chains *can* be divided into two categories: *discrete-time* and *continuous-time*. However, as we will see shortly, continuous-time chains implicitly involve discrete-time chains, and many authors do not bother to make the distinction. We will adopt this practice in situations where the distinction is not important. A discrete-time, discrete-valued Markov process will have a countable (not necessarily finite—just countable) number of possible outcomes. In general, the process will exhibit new outcomes at uniform intervals since a new outcome will be exhibited each time the discrete time index is incremented. Such a process is called a *discrete-time Markov chain*.

A continuous-time, discrete-valued Markov process will have a countable number of different possible outcomes, with transitions to new outcomes occurring at random points t_n in *continuous* time. However, if the outcomes, $\mathbf{x}(t_n^+)$, are considered as a function of n rather than of t_n, the result is a discrete-time Markov chain which is said to be *imbedded* in the continuous-time chain. [Note that $\mathbf{x}(t_n^+)$ represents the outcome just *after* the transition at t_n, and that $\mathbf{x}(t_n^-)$ represents the outcome just *before* the transition.]

In discussions of Markov chains, outcomes of the process are usually referred to as *states*. A discrete-time Markov chain can be specified in terms of its *state probabilities* and *state transition probabilities*. The *state probability* $p_i[n]$ is the probability that the chain will be in state i at time n.

$$p_i(n) = P\{s(n) = s_i\}$$

Usually the state probabilities for a Markov chain are given as *initial state probabilities* $p_i(0)$ which are the absolute probabilities that the chain will start in state i at time zero. The *transition probability* $p_{ij}(n_1, n_2)$ is the probability that the chain will enter state j at time n_2 given that the chain is in state i at time n_1.

$$p_{ij}(n_1, n_2) \equiv P\{s(n_2) = s_j \mid s(n_1) = s_i\}$$

In cases where $n_2 = n_1 + 1$, the transition probability is called the *one-step transition probability* and is usually denoted as $p_{ij}(n_2)$. In common usage the unqualified term *transition probability* refers to one-step transition probabilities, and the term *n-step transition probability* distinguishes the multistep case. The transition probabilities of a Markov chain are often represented as a stochastic matrix called the *determining matrix* of the chain.

Classification of Markov chains

Simple discrete-time Markov chains are chains in which the probability distribution of states at time k is fully determined by the state of the chain at the single instant of time $k - 1$. *Complex* discrete-time

Markov chains are chains in which the probability distribution of states at time k depends upon the states of the chain at two or more time instants prior to time k.

A *homogeneous Markov chain* is a Markov chain in which the transition probabilities depend only upon the difference between n_2 and n_1, rather than on the actual values of n_1 and n_2 themselves. (In other words, the transition probabilities are invariant under a shift of the origin.)

$$p_{ij}(n_1, n_2) = p_{ij}(n + n_2 - n_1) \qquad \text{for } n, n_1, n_2 \in \{0, 1, 2, \ldots\}$$

Most of the elementary literature deals primarily with homogeneous chains rather than with the more general case. Reference 11 refers to homogeneous chains as "chains with stationary probabilities p_{ij}."

A *stationary* Markov chain is a homogeneous Markov chain in which the state probabilities as well as the transition probabilities are invariant under a shift of the time origin.

$$[p_{ij}(n_1, n_2) = p_{ij}(n + n_2 - n_1)] \wedge [p_i(n) = p_i(0)]$$

$$\text{for all } n, n_1, n_2 \in \{0, 1, 2, \ldots\}$$

Thus the initial state probabilities equal the final state probabilities.

A Markov chain C_n with determining matrix **P** is called *regular* when the maximal eigenvalue of **P** (*remember*: $\lambda_{\max} = 1$ since **P** is stochastic) is a simple root of $P(\lambda)$ and all other eigenvalues of **P** have magnitudes strictly less than 1. If any of the final state probabilities p_j for a regular Markov chain C_n are equal to zero, the chain is described as *nonnegatively regular*. If all p_j are nonzero, then C_n is described as *positively regular*, or *normal*. If all the p_j are equal ($p_j = 1/n$), the chain is described as *completely regular*.

The vast majority of existing Markov chain theory deals with simple homogeneous discrete-time Markov chains with finite numbers of states. Kolmogorov calls such chains "Markov chains in the restricted sense of the word" [Ref. 12]. For brevity, C_n will be used in the remainder of this chapter to denote a simple homogeneous discrete-time Markov chain with n states.

State diagrams of Markov chains

The relationships between the various states and transition probabilities in a homogeneous Markov chain are often depicted as a state diagram. The transition probability from state A to state B is depicted as a directed path from state A to state B. State diagrams could be used to depict nonhomogeneous Markov chains, but this may become quite cumbersome, since in general a different set of transition probabilities will be needed for each time n.

Since each state must always have a next state, the probabilities for all paths exiting a state must sum to unity:

$$\forall i, \sum_j p_{ij} = 1$$

Often the probabilities for all the paths exiting a state as shown in a diagram will sum to something less than 1. In these cases there is an assumed or implied path which leaves and then immediately reenters the same state. The transition probability associated with this implied path is equal to the difference between unity and the sum of all the other exiting transition probabilities. If there are no other exiting pathways, the state is called an *absorbing state* since the chain will become trapped or absorbed in the state.

Example Consider the Markov chain depicted in Fig. 5.2. Inspection of the state diagram reveals that there must be an implied loop at state C having a transition probability, $p_{CC} = 0.15$. There are also implied loops at D ($p_{DD} = 1.0$) and at E ($p_{EE} = 0.5$). State D is an absorbing state. If we assign $s_1 \equiv A$, $s_2 \equiv B$, $s_3 \equiv C$ and so on, then we can represent the depicted Markov chain via the following stochastic matrix:

$$\begin{bmatrix} 0 & 0.30 & 0.25 & 0 & 0.45 & 0 \\ 0 & 0 & 0 & 0.57 & 0.43 & 0 \\ 0 & 0 & 0.15 & 0 & 0.20 & 0.65 \\ 0 & 0 & 0 & 1.00 & 0 & 0 \\ 0 & 0 & 0 & 0.50 & 0.50 & 0 \\ 0.79 & 0 & 0 & 0 & 0.21 & 0 \end{bmatrix} \quad (5.15)$$

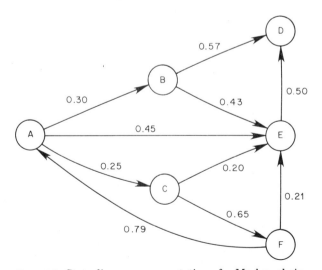

Figure 5.2 State diagram representation of a Markov chain.

Multistep transition probabilities

In a Markov chain, the probability of a multistep transition from one state to another is equal to the product of the transition probabilities for each of the single-step transitions which make up the multistep transition. The n-step transition probability (for a homogeneous chain) from state i to state j is often denoted as $p_{ij}^{(n)}$. Some authors, such as Karlin [Ref. 13] omit the parentheses on the superscript, but this notation can be confusing.

If the single-step transition probabilities are represented by the stochastic matrix $\mathbf{P}$, then all possible n-step transition probabilities are represented as a stochastic matrix $\mathbf{Q}$ given as:

$$\mathbf{Q} = \mathbf{P}^n = \underbrace{\mathbf{P} \cdot \mathbf{P} \cdot \mathbf{P} \cdots \cdot \mathbf{P}}_{n \text{ times}} \tag{5.16}$$

where the indicated multiplications are the "row-by-column" multiplications usually associated with matrices. Since the right-hand side of (5.16) can be factored into two powers of $\mathbf{P}$, we can write

$$\mathbf{P}^n = \mathbf{P}^m \mathbf{P}^{n-m} \tag{5.17}$$

Invoking the definition of matrix multiplication, we can express (5.17) as summations of matrix elements with explicit subscripts:

$$p_{ij}^{(n)} = \sum_{k=0}^{N} p_{ik}^{(m)} p_{kj}^{(n-m)} \quad \begin{array}{l} i = 0, 1, \ldots, N \\ j = 0, 1, \ldots, N \\ m = 0, 1, \ldots, m \end{array} \tag{5.18}$$

These summations are known as the *Chapman-Kolmogorov equations*. According to Ref. 11, these equations establish the connection between Markov chains and the theory of *semigroups* as discussed in Ref. 14.

Example Referring to Fig. 5.2, we conclude that the probabilities of a two-step transition from state A to state D is given by

$$p_{AD}^{(2)} = p_{AB}p_{BD} + p_{AE}p_{ED}$$
$$= (0.30)(0.57) + (0.45)(0.50) = 0.396$$

If we form the matrix of two-step transition probabilities,

$$\begin{bmatrix} 0 & 0 & 0.0375 & 0.396 & 0.404 & 0.1625 \\ 0 & 0 & 0 & 0.75 & 0.215 & 0 \\ 0.5135 & 0 & 0.0225 & 0.1 & 0.2665 & 0.0975 \\ 0 & 0 & 0 & 1.00 & 0 & 0 \\ 0 & 0 & 0 & 0.75 & 0.25 & 0 \\ 0 & 0.237 & 0.1975 & 0.105 & 0.4605 & 0 \end{bmatrix}$$

we see that the entry in row 1, column 4, is in fact 0.396. (Row 1 corresponds to starting state A, and column 4 corresponds to ending state D.)

Absolute state probabilities

For homogeneous Markov chains the absolute state probabilities and transition probabilities are related by

$$P_j(n) = \sum_{i=0}^{\infty} p_i(0) p_{ij}^{(n)}$$

where $P_j(n)$ = probability of chain being in a state j at time n
$P_{ij}^{(n)}$ = the n-step transition probability from i to j

First passage time [Ref. 15]

The *first passage time* from i to j is the time required for a Markov chain in state i to first reach state j. If $i = j$, the first passage time is called the *recurrence time*, or *first return time*, for state i, and it equals the time needed for the process to return to state i. In general, the first passage time will be a random variable, and we can let $f_{ij}^{(n)}$ represent the probability that the first passage time from i to j is equal to n. The values of $f_{ij}^{(n)}$ will be equal to the n-step transition probability from i to j minus the probabilities of all transitions from i to j which reach j in less than n steps and then visit other states before finally returning to j at step n. These probabilities must be subtracted because they are associated with a shorter first passage time determined by the *first* arrival at state j. Thus

$$f_{ij}^{(1)} = p_{ij}^{(1)}$$
$$f_{ij}^{(2)} = p_{ij}^{(2)} - f_{ij}^{(1)} p_{jj}^{(1)}$$
$$f_{ij}^{(3)} = p_{ij}^{(3)} - f_{ij}^{(1)} p_{jj}^{(2)} - f_{ij}^{(2)} p_{jj}^{(1)} \quad (5.19)$$
$$\vdots$$
$$f_{ij}^{(n)} = p_{ij}^{(n)} - f_{ij}^{(1)} p_{jj}^{(n-1)} - f_{ij}^{(2)} p_{jj}^{(n-2)} - \cdots - f_{ij}^{(n-1)} p_{jj}^{(1)}$$

The expected value μ_{ij} of the first passage time is given by

$$\mu_{ij} = 1 + \sum_{n=1}^{\infty} n f_{ij}^{(n)} \quad (5.20)$$

Furthermore, it can be shown that

$$\mu_{ij} = 1 + \sum_{k \neq j} p_{ik} \mu_{kj} \quad (5.21)$$

For a recurrent state i, we can obtain the mean recurrence time from (5.21) by setting $i = j$. If $\mu_{ii} < \infty$, then state i is called *positive recurrent*. If $\mu_{ii} = \infty$, then state i is called *null recurrent*. A finite state Markov chain will never contain null recurrent states.

Classification of states

The absolute probability that a given chain will eventually reach state j from state i can be obtained by summing the first passage probability over all possible passage times.

$$P\{j \text{ can be reached from } i\} = \sum_{n=1}^{\infty} f_{ij}^{(n)} \qquad (5.22)$$

If this probability is nonzero for a given i and j, then state j is said to be *accessible* from state i. If state j is accessible from state i and state i is accessible from state j, then we can say states i and j *communicate*. Communication is an equivalence relation. Since (5.22) is a probability, it must be less than or equal to unity. Thus for $i = j$, we have the probability that the chain will return to state i given by

$$\sum_{n=1}^{\infty} f_{ii}^{(n)} \leq 1 \qquad (5.23)$$

If the equality in (5.23) holds, it is a certainty that the chain will return to state i. In this case state i is a *recurrent* state. (In a finite state Markov chain, all expected recurrence times are finite, so all recurrent states are also *positive recurrent states*.) If the equality in (5.23) does not hold, state i is called a *transient* state. If the series

$$\sum_{n=1}^{\infty} p_{ij}^{(n)} \qquad (5.24)$$

converges, then state j is a transient state.

A positive recurrent state is also *periodic with period* τ if and only if all possible recurrence times share τ as their greatest common divisor ($\tau > 1$). If states i and j communicate and state i is periodic with period τ, then state j is also periodic with period τ. An *ergodic* state is a state which is both positive recurrent and aperiodic.

Reducibility of Markov chains

We can partition the state space of a Markov chain into classes or disjoint subsets such that all the states within each class communicate with each other but none of the states within a class communicate with any state in a different class. If states within a chain communicate with each of the other states, there will be only one class (which covers the entire state space), and the chain is called *irreducible*. (Some authors use the terms *nondecomposable* or *indecomposable*.)

For a *finite state* Markov chain, all states within any given equivalence class will either be all transient or all recurrent. A class containing transient states is called a *transient, inessential,* or *nonisolated group*. A class containing recurrent states is called a *recurrent, essential,* or *isolated group*.

Once a Markov chain enters a state which is in a recurrent class, it cannot subsequently enter a state outside of this class. The reasoning to prove this is simple—If a chain leaves a recurrent class to enter a different class, it can never return, and the ability to return is essential for the first class to be recurrent. If the chain could leave the first class and later return, the two classes would communicate and thus by definition could not be *different* classes.

The determining matrix **P** of a reducible Markov chain C_n can be decomposed such that the submatrices within **P** correspond to the equivalence classes which form the partition on C_n. An *ergodic* Markov chain is an irreducible chain whose states are all ergodic.

Final state probabilities

As the number of steps in a Markov chain becomes very large, the probability of ending up in state j can be determined in one of several ways depending upon the nature of state j and the starting state i.

- If j is a transient state, then

$$\lim_{n \to \infty} P_{ij}^{(n)} = 0$$

- If i and j are states within the same aperiodic recurrence class, then

$$\lim_{n \to \infty} P_{ij}^{(n)} = \pi_j = \frac{1}{\sum_{n=0}^{\infty} n f_{jj}^{(n)}}$$

where $f_{jj}^{(n)} \equiv 0$ and $p_{ii}^{(0)} \equiv 1$. This relationship is known as the *basic limit theorem of Markov chains*. In a finite state Markov chain, $\pi_j > 0$ for all j in the class, and the class is called *positive recurrent*, or *strongly ergodic*.

- If i and j are states within the same periodic recurrence class, then

$$\lim_{n \to \infty} P_{ij}^{(n)} = \pi_j = \frac{1}{n} \sum_{m=1}^{n} p_{ij}^{(m)}$$

- The final state probabilities p_j for a *regular* Markov chain C_n are given by

$$p_j = \frac{M_{jj}(1)}{\sum_{i=1}^{n} M_{ii}(1)} \qquad j = 1, 2, \ldots, n$$

where $M_{ii}(1)$ are the principal minors of $(\mathbf{P} - \lambda \mathbf{I})$ evaluated at $\lambda = 1$.

5.8 REFERENCES

1. H. Taub and D. L. Schilling: *Principles of Communication Systems*, 2d ed., McGraw-Hill, New York, 1986.
2. H. Urkowitz: *Signal Theory and Random Processes*, Artech House, Dedham, Mass., 1983.
3. A. Papoulis: *Probability, Random Variables, and Stochastic Processes*, 2d ed., McGraw-Hill, New York, 1984.
4. G. M. Jenkins and D. G. Watts: *Spectral Analysis and Its Applications*, Holden-Day, San Francisco, 1968.
5. S. Haykin: *Communication Systems*, 2d ed., Wiley, New York, 1983.
6. S. Stein and J. J. Jones: *Modern Communication Principles*, McGraw-Hill, New York, 1967.
7. R. S. Simpson and R. C. Houts: *Fundamentals of Analog and Digital Communication Systems*, Allyn and Bacon, Boston, 1971.
8. M. M. Blachman: *Noise and Its Effect on Communication*, McGraw-Hill, New York, 1966.
9. A. D. Whalen: *Detection of Signals in Noise*, Academic Press, New York, 1971.
10. A. B. Carlson: *Communication Systems: An Introduction to Signals and Noise in Electrical Communication*, McGraw-Hill, New York, 1968.
11. A. T. Bharucha-Reid: *Elements of the Theory of Markov Processes and Their Applications*, McGraw-Hill, New York, 1960.
12. V. I. Romanovsky: *Discrete Markov Chains*, Volters-Noordhoff Publishing, The Netherlands, 1970.
13. S. Karlin and H. M. Taylor: *A First Course in Stochastic Processes*, 2d ed., Academic Press, New York, 1975.
14. E. Hille and R. S. Phillips: *Functional Analysis and Semi-Groups*, American Mathematical Society, New York, 1957.
15. W. W. Hines and D. C. Montgomery: *Probability and Statistics in Engineering and Management Science*, 2d ed., Wiley, New York, 1980.
16. J. G. Kemeny and J. L. Snell: *Finite Markov Chains*, Springer-Verlag, New York, 1976.

Chapter 6

Signals and Spectra

6.1 Mathematical Modeling of Signals

Electronic signals are complicated phenomena whose exact behavior is impossible to describe completely. However, simple mathematical models can describe signals well enough to yield some very useful results in a variety of practical situations. The distinction between a signal and its mathematical representation is not always rigidly observed in the literature. Mathematical functions which only model signals are frequently referred to as *signals*, and properties of these functions are often taken as properties of the signals themselves.

Mathematical models of signals can be categorized as either *steady-state* or *transient*. The typical voltage output from an oscillator is sketched in Fig. 6.1. This signal exhibits three different parts—a *turn-on transient* at the beginning, an interval of *steady-state* operation in the middle, and a *turn-off transient* at the end. It is possible to formulate a single mathematical expression that describes all three parts, but for most uses, such an expression would be unnecessarily complicated. In cases where the primary concern is steady-state behavior, simplified mathematical expressions that ignore the transients will often be adequate. The steady-state portion of the oscillator output can be modeled as a sinusoid that theoretically exists for all time. The transients might be modeled as exponentially saturating and decaying sinusoids as shown in Figs. 6.2 and 6.3. In Fig. 6.2, the saturating exponential envelope continues to increase, but it never quite reaches the steady-state value. Likewise, the decaying exponential envelope in Fig. 6.3 continues to decrease, but it never quite reaches zero. In this context, the steady-state value is sometimes called an *asymptote*, or the envelope can be said to *asymptotically* approach the steady-state value.

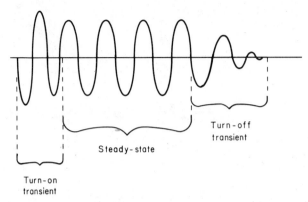

Figure 6.1 Typical output from an audio oscillator.

Steady-state and transient models of signal behavior inherently contradict each other, and neither constitutes a "true" description of a particular signal. The formulation of an appropriate model requires an understanding of the signal to be modeled and of the implications that a particular choice of model will have for the intended application.

The following definitions and concepts will prove useful in discussions of functions used to model steady-state signals.

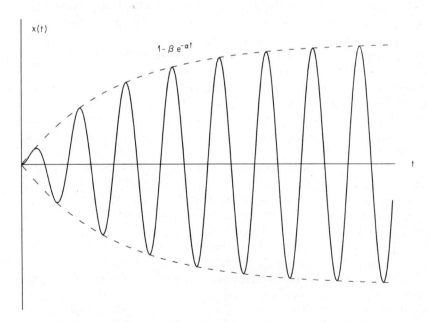

Figure 6.2 Exponentially saturating sinusoid.

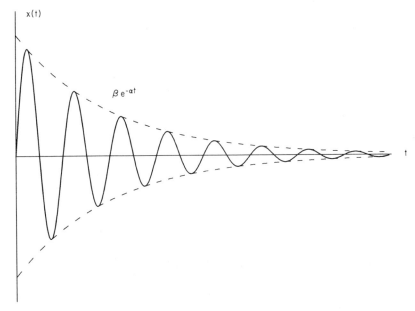

Figure 6.3 Exponentially decaying sinusoid.

Definition. A function $x(t)$ is periodic with a period of T if and only if $x(t + nT) = x(t)$ for all integer values of n.

Definition. A function $x(t)$ is said to be *even*, or to exhibit *even symmetry*, if, for all t, $x(t) = x(-t)$ such as shown in Fig. 6.4.

Definition. A function $x(t)$ is said to be *odd*, or to exhibit *odd symmetry*, if, for all t, $x(t) = -x(-t)$ such as shown in Fig. 6.5.

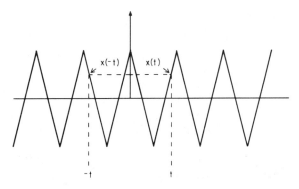

Figure 6.4 Even-symmetric function.

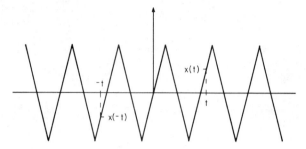

Figure 6.5 Odd-symmetric function.

Addition and multiplication of symmetric functions will obey the following rules:

$$\text{Even} + \text{even} = \text{even} \tag{6.1}$$

$$\text{Odd} + \text{odd} = \text{odd} \tag{6.2}$$

$$\text{Odd} \times \text{odd} = \text{even} \tag{6.3}$$

$$\text{Even} \times \text{even} = \text{even} \tag{6.4}$$

$$\text{Odd} \times \text{even} = \text{odd} \tag{6.5}$$

Any *periodic* function can be resolved into a sum of an even function and an odd function as given by

$$x(t) = x_{\text{even}}(t) + x_{\text{odd}}(t) \tag{6.6}$$

$$x_{\text{even}}(t) = \frac{1}{2}[x(t) + x(-t)] \tag{6.7}$$

$$x_{\text{odd}}(t) = \frac{1}{2}[x(t) - x(-t)] \tag{6.8}$$

6.2 Energy Signals versus Power Signals

It is a common practice to deal with mathematical functions representing abstract signals as though they are either voltages across a 1-Ω resistor or currents through a 1-Ω resistor. Since, in either case, the resistance has an assumed value of unity, the voltage and current for any particular signal will be numerically equal—thus obviating the need to select one viewpoint over the other. Thus for a signal $x(t)$, the instantaneous power $p(t)$ dissipated in the 1-Ω resistor is simply the squared amplitude of the signal

$$p(t) = |x(t)|^2 \tag{6.9}$$

regardless of whether $x(t)$ represents a voltage or a current. To emphasize the fact that the power given by (6.9) is based upon unity resistance, it is often referred to as the *normalized power*. The total energy of the signal $x(t)$ is then obtained by integrating the right-hand side (RHS) of (6.9) over all time,

$$E = \int_{-\infty}^{\infty} |x(t)|^2 \, dt \tag{6.10}$$

and the average power is given by

$$P = \lim_{T \to \infty} \frac{1}{T} \int_{-T/2}^{T/2} |x(t)|^2 \, dt \tag{6.11}$$

A few texts (for example, Ref. 1) equivalently define the average power as:

$$P = \lim_{T \to \infty} \frac{1}{2T} \int_{-T}^{T} |x(t)|^2 \, dt \tag{6.12}$$

If the total energy is finite and nonzero, $x(t)$ is referred to as an *energy signal*. If the average power is finite and nonzero, $x(t)$ is referred to as a *power signal*. Note that a power signal has infinite energy and an energy signal has zero average power; thus the two categories are mutually exclusive. Periodic signals and most random signals are power signals, while most deterministic aperiodic signals are energy signals.

6.3 Fourier Series

Trigonometric forms

Periodic signals can be resolved into linear combinations of phase-shifted sinusoids using the *Fourier series*, which is given by

$$x(t) = \frac{a_0}{2} + \sum_{n=1}^{\infty} [a_n \cos(n\omega_0 t) + b_n \sin(n\omega_0 t)] \tag{6.13}$$

where $a_0 = \frac{2}{T} \int_{-T/2}^{T/2} x(t) \, dt \tag{6.14}$

$a_n = \frac{2}{T} \int_{-T/2}^{T/2} x(t) \cos(n\omega_0 t) \, dt \tag{6.15}$

$b_n = \frac{2}{T} \int_{-T/2}^{T/2} x(t) \sin(n\omega_0 t) \, dt \tag{6.16}$

$T =$ period of $x(t)$

$\omega_0 = \frac{2\pi}{T} = 2\pi f_0 =$ fundamental radian frequency of $x(t)$

Upon application of the appropriate trigonometric identities, Eq. (6.13) can be put into the following alternative form:

$$x(t) = c_0 + \sum_{n=1}^{\infty} c_n \cos(n\omega_0 t - \theta_n) \qquad (6.17)$$

where the c_n and θ_n are obtained from a_n and b_n using

$$c_0 = \frac{a_0}{2} \qquad (6.18)$$

$$c_n = \sqrt{a_n^2 + b_n^2} \qquad (6.19)$$

$$\theta_n = \tan^{-1}\left(\frac{b_n}{a_n}\right) \qquad (6.20)$$

Examination of (6.13) and (6.17) reveals that a periodic signal contains only a dc component plus sinusoids whose frequencies are integer multiples of the original signal's *fundamental* frequency. (For a fundamental frequency of f_0, $2f_0$ is the *second harmonic*, $3f_0$ is the *third harmonic*, etc.) Theoretically, periodic signals will generally contain an infinite number of harmonically related sinusoidal components. In the real world, however, periodic signals will contain only a finite number of measurable harmonics. Consequently, pure mathematical functions are only approximately equal to the practical signals which they model.

Exponential form

The trigonometric form of the Fourier series given by (6.13) makes it easy to visualize periodic signals as summations of sine and cosine waves, but mathematical manipulations are often more convenient when the series is in the exponential form given by

$$x(t) = \sum_{n=-\infty}^{\infty} c_n e^{j2\pi n f_0 t} \qquad (6.21)$$

where $c_n = \dfrac{1}{T} \displaystyle\int_T x(t) e^{-j2\pi n f_0 t}\, dt \qquad (6.22)$

The integral notation used in (6.22) indicates that the integral is to be evaluated over one period of $x(t)$. In general, the values of c_n are complex; and they are often presented in the form of magnitude and phase spectra as shown in Fig. 6.6. The magnitude and phase values

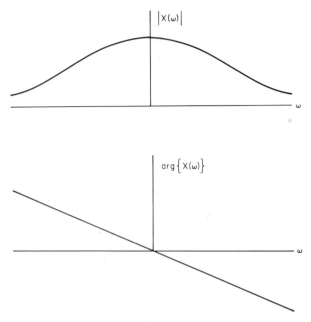

Figure 6.6 Magnitude and phase spectra.

plotted in such spectra are obtained from the c_n using

$$|c_n| = \sqrt{(Re\{c_n\})^2 + (Im\{c_n\})^2} \qquad (6.23)$$

$$\theta_n = \tan^{-1}\left(\frac{Im\{c_n\}}{Re\{c_n\}}\right) \qquad (6.24)$$

The complex c_n of (6.22) can be obtained from the a_n and b_n of (6.11) and (6.12) using

$$c_n = \begin{cases} \dfrac{a_n + jb_n}{2} & n < 0 \\ a_0 & n = 0 \\ \dfrac{a_n - jb_n}{2} & n > 0 \end{cases} \qquad (6.25)$$

Conditions of applicability

The Fourier series can be applied to almost all periodic signals of *practical* interest. However, there are some functions for which the series will not converge. The Fourier series coefficients are guaranteed to exist, and the series will converge uniformly, if $x(t)$ satisfies the following conditions:

1. The function $x(t)$ is a single-valued function.
2. The function $x(t)$ has at most a finite number of discontinuities within each period.
3. The function $x(t)$ has at most a finite number of extrema (i.e., maxima and minima) within each period.
4. The function $x(t)$ is absolutely integrable over a period:

$$\int_T |x(t)|\, dt < \infty \qquad (6.26)$$

These conditions are often called the *Dirichlet conditions* in honor of Peter Gustav Lejeune Dirichlet (1805–1859) who first published them in the 1828 issue of *Journal für die reine und angewandte Mathematik* (commonly known as *Crelle's Journal*). In applications where it is sufficient for the Fourier series coefficients to be convergent in the mean, rather than uniformly convergent, it suffices for $x(t)$ to be integrable-square over a period:

$$\int_T |x(t)|^2\, dt < \infty \qquad (6.27)$$

For most engineering purposes, the Fourier series is usually assumed to be identical to $x(t)$ if conditions 1 through 3 plus either (6.26) or (6.27) are satisfied.

Properties of the Fourier series

A number of useful Fourier series properties are listed in Table 6.1. For ease of notation, the coefficients c_n corresponding to $x(t)$ are denoted as $X(n)$, and the c_n corresponding to $y(t)$ are denoted as $Y(n)$. In other words, the Fourier series (FS) representations of $x(t)$ and $y(t)$ are given by

$$x(t) = \sum_{n=-\infty}^{\infty} X(n) \exp\left(\frac{j2\pi nt}{T}\right) \qquad (6.28)$$

$$y(t) = \sum_{n=-\infty}^{\infty} Y(n) \exp\left(\frac{j2\pi nt}{T}\right) \qquad (6.29)$$

where T is the period of both $x(t)$ and $y(t)$. In addition to the properties listed in Table 6.1, the Fourier series coefficients exhibit certain symmetries. If (and only if) $x(t)$ is real, the corresponding FS

TABLE 6.1 Properties of the Fourier Series

Property	Time function	Transform
1. Homogeneity	$ax(t)$	$aX(n)$
2. Additivity	$x(t) + y(t)$	$X(n) + Y(n)$
3. Linearity	$ax(t) + by(t)$	$aX(n) + bY(n)$
4. Multiplication	$x(t)y(t)$	$\sum_{m=-\infty}^{\infty} X(n-m)Y(m)$
5. Convolution	$\dfrac{1}{T}\int_0^T x(t-\tau)y(\tau)\,d\tau$	$X(n)Y(n)$
6. Time shifting	$x(t-\tau)$	$\exp\!\left(\dfrac{-j2\pi n\tau}{T}\right)X(n)$
7. Frequency shifting	$\exp\!\left(\dfrac{j2\pi mt}{T}\right)x(t)$	$X(n-m)$

Note: $x(t)$, $y(t)$, $X(n)$, and $Y(n)$ are as given in Eqs. (6.28) and (6.29).

coefficients will exhibit even symmetry in their real part and odd symmetry in their imaginary part.

$$Im[x(t)] = 0 \Leftrightarrow Re[X(-n)] = Re[X(n)]$$
$$Im[X(-n)] = Im[X(n)] \quad (6.30)$$

Equation (6.30) can be rewritten in a more compact form as:

$$Im[x(t)] = 0 \Leftrightarrow X(-n) = X^*(n) \quad (6.31)$$

where the superscript asterisk indicates complex conjugation. Likewise for purely imaginary $x(t)$, the corresponding FS coefficients will exhibit odd symmetry in the real part and even symmetry in their imaginary part:

$$Re[x(t)] = 0 \Leftrightarrow X(-n) = -[X^*(n)] \quad (6.32)$$

If and only if $x(t)$ is (in general) complex with even symmetry in the real part and odd symmetry in the imaginary part, then the corresponding FS coefficients will be purely real:

$$x(-t) = x^*(t) \Leftrightarrow Im[X(n)] = 0 \quad (6.33)$$

If and only if $x(t)$ is (in general) complex with odd symmetry in the real part and even symmetry in the imaginary part, then the corresponding FS coefficients will be purely imaginary:

$$x(-t) = -[x^*(t)] \Leftrightarrow Re[X(n)] = 0 \tag{6.34}$$

In terms of the amplitude and phase spectra, Eq. (6.31) means that for real signals, the amplitude spectrum will have even symmetry and the phase spectrum will have odd symmetry. If $x(t)$ is both real and even, then both (6.31) and (6.33) apply. In this special case, the FS coefficients will be both real and even-symmetric. At first glance, it may appear that real-even-symmetric coefficients are in contradiction to the expected odd-symmetric phase spectrum; but in fact there is no contradiction. For all the positive real coefficients, the corresponding phase is of course zero. For each of the negative real coefficients, we can choose a phase value of either plus or minus 180°. By appropriate selection of positive and negative values, odd symmetry in the phase spectrum can be maintained.

Fourier series of a square wave

Consider the square wave shown in Fig. 6.7. The Fourier series representation of this signal is given by

$$x(t) = \sum_{n=-\infty}^{\infty} c_n \exp\left(\frac{j2\pi nt}{T}\right) \tag{6.35}$$

where $c_n = \dfrac{\tau A}{T} \operatorname{sinc}\left(\dfrac{n\tau}{T}\right)$ \hfill (6.36)

Since the signal is both real and even-symmetric, the FS coefficients are real and even-symmetric as shown in Fig. 6.8. The corresponding amplitude spectrum will be even as shown in Fig. 6.9a. Appropriate selection of $\pm 180°$ values for the phase of negative coefficients will allow an odd-symmetric phase spectrum to be plotted as in Fig. 6.9b.

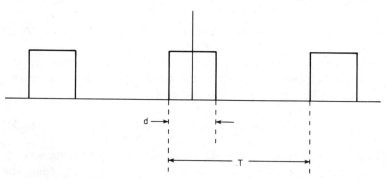

Figure 6.7 Square wave.

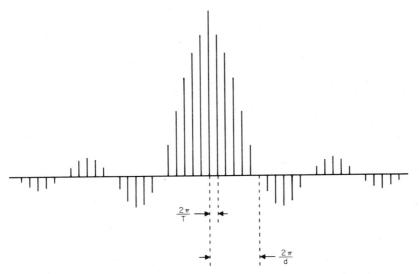

Figure 6.8 Fourier series for a square wave.

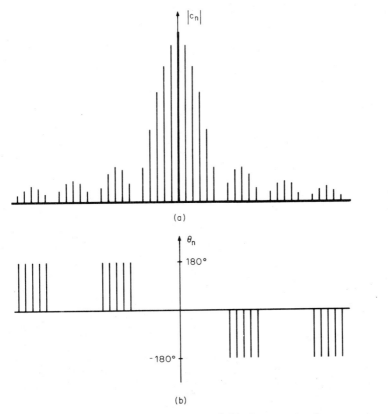

Figure 6.9 Fourier series (a) amplitude and (b) phase spectra for a square wave.

Parseval's theorem

The average power (normalized for 1 Ω) of a real-valued periodic function of time can be obtained directly from the Fourier series coefficients by using Parseval's theorem:

$$P = \frac{1}{T} \int_T |x(t)|^2 \, dt$$

$$= \sum_{n=-\infty}^{\infty} |c_n|^2 = c_0^2 + \sum_{n=1}^{\infty} \frac{1}{2} |2c_n|^2 \qquad (6.37)$$

6.4 Fourier Transform

The Fourier transform is defined as:

$$X(f) = \int_{-\infty}^{\infty} x(t) \, e^{-j2\pi ft} \, dt \qquad (6.38)$$

and the inverse transform is defined as:

$$x(t) = \int_{-\infty}^{\infty} X(f) \, e^{j2\pi ft} \, df \qquad (6.39)$$

There are a number of different shorthand notations for indicating that $x(t)$ and $X(f)$ are related via the Fourier transform. Some of the more common notations include:

$$X(f) = \mathscr{F}[x(t)] \qquad (6.40)$$

$$x(t) = \mathscr{F}^{-1}[X(f)] \qquad (6.41)$$

$$x(t) \overset{\text{FT}}{\leftrightarrow} X(f) \qquad (6.42)$$

$$x(t) \underset{\text{IFT}}{\overset{\text{FT}}{\rightleftarrows}} X(f) \qquad (6.43)$$

$$x(t) \qquad X(f) \qquad (6.44)$$

The notation used in (6.40) and (6.41) is easiest to typeset, while the notation of (6.44) is probably the most difficult. However, the notation of (6.44) is used in the classic work on fast Fourier transforms by Brigham [Ref. 2]. The notations of (6.42) and (6.43) while more difficult to typeset, offer the flexibility of changing the letters FT to FS, DFT, or DTFT to indicate, respectively, "Fourier series," "discrete Fourier transform," or "discrete-time Fourier transform," as is done in Ref. 3. (The latter two transforms will be discussed in Chap. 11.) The form used in (6.43) is perhaps best saved for tutorial situations (such as Ref. 4) where the distinction between the transform and inverse transform needs to be emphasized. Strictly speaking, the equality shown in (6.41) is incorrect, since the inverse transform of $X(f)$ is only guaranteed to approach $x(t)$ in the sense of convergence in the mean. Nevertheless, the notation of Eq. (6.41) appears often throughout the engineering literature. A number of useful Fourier transform properties are listed in Table 6.2.

TABLE 6.2 Properties of the Fourier Transform

Property	Time function	Transform
	$x(t)$	$X(f)$
1. Homogeneity	$ax(t)$	$aX(f)$
2. Additivity	$x(t) + y(t)$	$X(f) + Y(f)$
3. Linearity	$ax(t) + by(t)$	$aX(f) + bY(f)$
4. Differentiation	$\dfrac{d^n}{dt^n} x(t)$	$(j2\pi f)^n X(f)$
5. Integration	$\int_{-\infty}^{t} x(\tau)\, d\tau$	$\dfrac{X(f)}{j2\pi f} + \dfrac{1}{2} X(0)\, \delta(f)$
6. Frequency shifting	$e^{-j2\pi f_0 t} x(t)$	$X(f + f_0)$
7. Sine modulation	$x(t) \sin(2\pi f_0 t)$	$\dfrac{1}{2}[X(f - f_0) + X(f + f_0)]$
8. Cosine modulation	$x(t) \cos(2\pi f_0 t)$	$\dfrac{1}{2}[X(f - f_0) - X(f + f_0)]$
9. Time shifting	$x(t - \tau)$	$e^{-j\omega\tau} X(f)$
10. Time convolution	$\int_{-\infty}^{\infty} h(t - \tau) x(\tau)\, d\tau$	$H(f) X(f)$
11. Multiplication	$x(t) y(t)$	$\int_{-\infty}^{\infty} X(\lambda) Y(f - \lambda)\, d\lambda$
12. Scaling time and frequency	$x\left(\dfrac{t}{a}\right) \quad a > 0$	$aX(af)$
13. Duality	$X(t)$	$x(-f)$
14. Conjugation	$x^*(t)$	$X^*(-f)$
15. Real part	$\mathrm{Re}[x(t)]$	$\dfrac{1}{2}[X(f) + X^*(-f)]$
16. Imaginary part	$\mathrm{Im}[x(t)]$	$\dfrac{1}{2j}[X(f) - X^*(-f)]$

Fourier transforms of periodic signals

Often there is a requirement to analyze systems that include both periodic power signals and aperiodic energy signals. The mixing of Fourier transform results and Fourier series results implied by such an analysis may be quite cumbersome. For the sake of convenience, the spectra of most periodic signals can be obtained as Fourier transforms that involve the Dirac delta function. When the spectrum of a periodic signal is determined via the Fourier series, the spectrum will consist of lines located at the fundamental frequency and its harmonics. When the spectrum of this same signal is obtained as a Fourier transform, the spectrum will consist of Dirac delta functions located at the fundamental frequency and its harmonics. Obviously, these two different mathematical representations must be equivalent in their physical significance. Specifically, consider a periodic signal $x_p(t)$ having a period of T. The Fourier series representation of $x_p(t)$ is obtained from Eq. (6.21) as:

$$x_p(t) = \sum_{n=-\infty}^{\infty} c_n \exp\left(\frac{j2\pi nt}{T}\right) \quad (6.45)$$

We can then define a *generating function* $x(t)$, which is equal to a single period of $x_p(t)$:

$$x(t) = \begin{cases} x_p(t) & |t| \leq \frac{T}{2} \\ 0 & \text{elsewhere} \end{cases} \quad (6.46)$$

The periodic signal $x_p(t)$ can be expressed as an infinite summation of time-shifted copies of $x(t)$:

$$x_p(t) = \sum_{n=-\infty}^{\infty} x(t - nT) \quad (6.47)$$

The Fourier series coefficients c_n appearing in (6.45) can be obtained and

$$c_n = \frac{1}{T} X\left(\frac{n}{T}\right) \quad (6.48)$$

where $X(f)$ is the Fourier transform of $x(t)$. Thus, the Fourier transform of $x_p(t)$ can be obtained as:

$$\mathscr{F}[x_p(t)] = \frac{1}{T} \sum_{n=-\infty}^{\infty} X\left(\frac{n}{T}\right) \delta\left(f - \frac{n}{T}\right) \quad (6.49)$$

Common Fourier transform pairs

A number of frequently encountered Fourier transform pairs are listed in Table 6.3. Several of these pairs are actually obtained as Fourier transforms-in-the-limit.

6.5 Spectral Density

Energy spectral density

The *energy spectral density* of an energy signal is defined as the squared magnitude of the signal's Fourier transform:

$$S_e(f) = |X(f)|^2 \qquad (6.50)$$

Analogous to the way in which Parseval's theorem relates the Fourier series coefficients to the average power of a power signal, *Rayleigh's energy theorem* relates the Fourier transform to the total energy of an energy signal as follows:

$$E = \int_{-\infty}^{\infty} |x(t)|^2 \, dt = \int_{-\infty}^{\infty} S_e(f) \, df = \int_{-\infty}^{\infty} |X(f)|^2 \, df \qquad (6.51)$$

In many texts where $x(t)$ is assumed to be real-valued, the absolute value signs are omitted from the first integrand in (6.51). In some texts (such as Ref. 8), Eq. (6.51) is loosely referred to as Parseval's theorem.

Power spectral density of a periodic signal

The *power spectral density* (psd) of a periodic signal is defined as the squared magnitude of the signal's line spectrum obtained via either a Fourier series or a Fourier transform with impulses. Using the Dirac delta notational conventions of the latter, the psd is defined as:

$$S_p(f) = \frac{1}{T^2} \sum_{n=-\infty}^{\infty} \delta\left(f - \frac{n}{T}\right) \left|X\left(\frac{n}{T}\right)\right|^2 \qquad (6.52)$$

where T is the period of the signal $x(t)$. Parseval's theorem as given by Eq. (6.37) can be restated in the notation of Fourier transform spectra as:

$$P = \frac{1}{T^2} \sum_{n=-\infty}^{\infty} \left|X\left(\frac{n}{T}\right)\right|^2 \qquad (6.53)$$

TABLE 6.3 Some Common Fourier Transform Pairs

Pair no.	$x(t)$	$X(\omega)$	$X(f)$
1	1	$2\pi\,\delta(\omega)$	$\delta(f)$
2	$u_1(t)$	$\dfrac{1}{j\omega} + \pi\,\delta(\omega)$	$\dfrac{1}{2\pi f} + \dfrac{1}{2}\delta(f)$
3	$\delta(t)$	1	1
4	t^n	$2\pi j^n\,\delta^{(n)}(\omega)$	$\left(\dfrac{j}{2\pi}\right)^n \delta^{(n)}(f)$
5	$\sin\omega_0 t$	$j\pi[\delta(\omega+\omega_0) - \delta(\omega-\omega_0)]$	$\dfrac{j}{2}[\delta(f+f_0) - \delta(f-f_0)]$
6	$\cos\omega_0 t$	$\pi[\delta(\omega+\omega_0) + \delta(\omega-\omega_0)]$	$\dfrac{1}{2}[\delta(f+f_0) + \delta(f-f_0)]$
7	$e^{-at}u_1(t)$	$\dfrac{1}{j\omega + a}$	$\dfrac{1}{j2\pi f + a}$

8	$u_1(t)e^{-at}\sin\omega_0 t$	$\dfrac{\omega_0}{(a+j\omega)^2+\omega_0^2}$	$\dfrac{2\pi f_0}{(a+j2\pi f)^2+(2\pi f_0)^2}$
9	$u_1(t)e^{-at}\cos\omega_0 t$	$\dfrac{a+j\omega}{(a+j\omega)^2+\omega_0^2}$	$\dfrac{a+j2\pi f}{(a+j2\pi f)^2+(2\pi f_0)^2}$
10	$\begin{cases}1 & \lvert t\rvert\leq\dfrac{1}{2}\\ 0 & \text{elsewhere}\end{cases}$	$\operatorname{sinc}\left(\dfrac{\omega}{2\pi}\right)$	$\operatorname{sinc} f$
11	$\operatorname{sinc} t \triangleq \dfrac{\sin\pi t}{\pi t}$	$\begin{cases}1 & \lvert\omega\rvert\leq\pi\\ 0 & \text{elsewhere}\end{cases}$	$\begin{cases}1 & \lvert f\rvert\leq\dfrac{1}{2}\\ 0 & \text{elsewhere}\end{cases}$
12	$\begin{cases}at\exp(-at) & t>0\\ 0 & \text{elsewhere}\end{cases}$	$\dfrac{a}{(a+j\omega)^2}$	$\dfrac{a}{(a+j2\pi f)^2}$
13	$\exp(-a\lvert t\rvert)$	$\dfrac{2a}{a^2+\omega^2}$	$\dfrac{2a}{a^2+4\pi^2 f^2}$
14	$\operatorname{signum} t \triangleq \begin{cases}1 & t>0\\ 0 & t=0\\ -1 & t<0\end{cases}$	$\dfrac{2}{j\omega}$	$\dfrac{1}{j\pi f}$

6.6 Autocorrelation

Autocorrelation of energy signals

Let $x(t)$ represent an energy signal that is, in general, complex-valued. The autocorrelation of $x(t)$ is given by

$$B_e(\tau) = \int_{-\infty}^{\infty} x(t)x^*(t-\tau)\,dt$$

$$= \int_{-\infty}^{\infty} x(t+\tau)x^*(t)\,dt \qquad (6.54)$$

The autocorrelation function thus defined is, in general, complex-valued with an even-symmetric real part and an odd-symmetric imaginary part.

$$Re[R_e(\tau)] = Re[R_e(-\tau)] \qquad (6.55)$$

$$Im[R_e(\tau)] = -Im[R_e(-\tau)] \qquad (6.56)$$

Equations (6.55) and (6.56) can be rewritten in a more compact form as:

$$R_e(\tau) = R_e^*(-\tau) \qquad (6.57)$$

When (6.54) is evaluated at $\tau = 0$, the result is equal to the total energy of the signal. This can be easily verified by substituting $\tau = 0$ into (6.54) to obtain

$$R_e(0) = \int_{-\infty}^{\infty} x(t)x^*(t)\,dt = \int_{-\infty}^{\infty} |x(t)|^2\,dt = E \qquad (6.58)$$

Furthermore, it can be shown that the maximum value of the autocorrelation function occurs at $\tau = 0$.

$$R_e(0) \geq |R_e(\tau)| \quad \text{for all } \tau$$

This makes sense when the autocorrelation is viewed as a measure of similarity between $x(t)$ and delayed copies of $x(t)$—the two can never be more similar than when there is zero delay and they exactly coincide. The autocorrelation defined by (6.54) and the energy spectral density defined by Eq. (6.50) form a Fourier transform pair:

$$R_e(\tau) \overset{FT}{\leftrightarrow} S_e(f) \qquad (6.59)$$

Autocorrelation of power signals

Let $x(t)$ represent a periodic power signal that is, in general, complex-valued. The autocorrelation of $x(t)$ is then given by

$$B_p(\tau) = \frac{1}{T} \int_{-T/2}^{T/2} x(t) x^*(t - \tau) \, dt \tag{6.60}$$

Just as for the case of an energy signal, the autocorrelation function defined by (6.60) will exhibit even symmetry in its real part and odd symmetry in its imaginary part.

$$R_p(\tau) = R_p^*(-\tau) \tag{6.61}$$

When (6.60) is evaluated at $\tau = 0$, the result is equal to the average power of the signal:

$$R_p(0) = \frac{1}{T} \int_{-T/2}^{T/2} |x(t)|^2 \, dt = P_{\text{avg}} \tag{6.62}$$

As we might by now expect, the maximum value of $R_p(\tau)$ occurs at $\tau = 0$:

$$R_p(0) \geq |R_p(\tau)| \qquad \text{for all } \tau$$

The autocorrelation defined by (6.60) and the power spectral density defined by Eq. (6.52) form a Fourier transform pair:

$$R_p(\tau) \overset{\text{FT}}{\leftrightarrow} S_p(f) \tag{6.63}$$

6.7 Cross-Correlations

Cross-correlation of energy signals

Let $x(t)$ and $y(t)$ represent a pair of energy signals that are, in general, complex-valued. Cross-correlations of x and y can be defined by

$$R_{xy}(\tau) = \int_{-\infty}^{\infty} x(t) y^*(t - \tau) \, dt \tag{6.64}$$

$$R_{yx}(\tau) = \int_{-\infty}^{\infty} y(t) x^*(t - \tau) \, dt \tag{6.65}$$

Notice that $R_{xy}(\tau)$ and $R_{yx}(\tau)$ are not equal; in fact,

$$R_{xy}(\tau) = R_{yx}^*(-\tau) \tag{6.66}$$

Furthermore, it can be shown that

$$R_{xy}(\tau) \overset{FT}{\leftrightarrow} X(f)Y^*(f) \tag{6.67}$$

$$R_{yx}(\tau) \overset{FT}{\leftrightarrow} Y(f)X^*(f) \tag{6.68}$$

Cross-correlation of power signals

Let $x(t)$ and $y(t)$ represent a pair of power signals that are, in general, complex-valued. Cross-correlations of x and y can be defined by

$$R_{xy}(\tau) = \lim_{T \to \infty} \frac{1}{2T} \int_{-T}^{T} x(t)y^*(t - \tau)\, dt \tag{6.69}$$

$$R_{yx}(\tau) = \lim_{T \to \infty} \frac{1}{2T} \int_{-T}^{T} y(t)x^*(t - \tau)\, dt \tag{6.70}$$

If $x(t)$ and $y(t)$ are each periodic with a period of T_0, the cross-correlations can be simplified to

$$R_{xy}(\tau) = \frac{1}{T_0} \int_{-T_0/2}^{T_0/2} x(t)y^*(t - \tau)\, dt \tag{6.71}$$

$$R_{yx}(\tau) = \frac{1}{T_0} \int_{-T_0/2}^{T_0/2} y(t)x^*(t - \tau)\, dt \tag{6.72}$$

It should be noted that in this case, both $R_{xy}(\tau)$ and $R_{yx}(\tau)$ will be periodic with a period of T_0. Furthermore, it can be shown that

$$R_{xy}(\tau) \overset{FT}{\leftrightarrow} \frac{1}{T_0^2} \sum_{n=-\infty}^{\infty} X\left(\frac{n}{T_0}\right) Y^*\left(\frac{n}{T_0}\right) \delta\left(f - \frac{n}{T_0}\right) \tag{6.73}$$

$$R_{yx}(\tau) \overset{FT}{\leftrightarrow} \frac{1}{T_0^2} \sum_{n=-\infty}^{\infty} Y\left(\frac{n}{T_0}\right) X^*\left(\frac{n}{T_0}\right) \delta\left(f - \frac{n}{T_0}\right) \tag{6.74}$$

6.8 Bandpass Signals [Refs. 1 and 5]

Several types of modulated signals are most conveniently represented as

$$x(t) = a(t) \cos[\omega_0 t + \phi(t)] \tag{6.75}$$

where $x(t)$ is real-valued. For amplitude modulation, $a(t)$ varies as the message varies while $\phi(t)$ is held constant. For phase modulation, $\phi(t)$ varies as the message varies while $a(t)$ is held constant. The component $a(t)$ is called the *envelope*, or *natural envelope*, of the signal $x(t)$. Other names sometimes given to $a(t)$ include *instantaneous amplitude*, *instantaneous envelope,* or *amplitude modulation.* The component $\phi(t)$ is called the *phase, phase modulation, phase deviation,* or *instantaneous phase deviation* of the signal $x(t)$. The signal representation of (6.75) can be expanded and put into the *quadrature form* given by

$$x(t) = x_c(t) \cos \omega_0 t - x_s(t) \sin \omega_0 t \tag{6.76}$$

The component $x_c(t)$ is called the *inphase modulation component* and is sometimes denoted as $I(t)$. The component $x_s(t)$ is called the *quadrature modulation component* and is sometimes denoted as $Q(t)$. The choice of reference frequency ω_0 can be somewhat arbitrary, but it is usually set equal to the actual carrier frequency exhibited by $x(t)$. Mathematically, any signal $x(t)$ could be put into the form of (6.76) but such a representation is most useful when $x_c(t)$ and $x_s(t)$ are lowpass signals. It can be shown [Ref. 5] that if the bandwidth of $x(t)$ is less than $2\omega_0$, then $x_c(t)$ and $x_s(t)$ will be lowpass signals given by

$$x_c(t) = a(t) \cos[\phi(t)] \tag{6.77}$$

$$x_s(t) = a(t) \sin[\phi(t)] \tag{6.78}$$

The envelope and phase of $x(t)$ can be obtained from the inphase and quadrature components by using

$$a(t) = \sqrt{x_c^2(t) + x_s^2(t)} \tag{6.79}$$

$$\phi(t) = \tan^{-1}\left[\frac{x_s(t)}{x_c(t)}\right] \tag{6.80}$$

Complex envelope

The complex envelope of $x(t)$ is defined as:

$$\tilde{x}(t) = x_c(t) + jx_s(t) = a(t) \exp[j\phi(t)] \tag{6.81}$$

The bandpass signal $x(t)$ can be expressed in terms of the complex envelope as:

$$x(t) = Re[\tilde{x}(t) \exp(j\omega_0 t)] \tag{6.82}$$

The validity of (6.82) can be demonstrated by substituting (6.77) and (6.78) into (6.82) and applying the appropriate trigonometric identities.

$$\begin{aligned} Re[\tilde{x}(t) \exp(j\omega_0 t)] &= Re\{[a(t) \cos[\phi(t)] + ja(t) \sin[\phi(t)]] \\ &\quad \cdot [\cos \omega_0 t + j \sin \omega_0 t]\} \\ &= a(t) \cos[\phi(t)] \cos(\omega_0 t) - a(t) \sin[\phi(t)] \sin(\omega_0 t) \\ &= a(t) \cos[\omega_0 t + \phi(t)] = x(t) \end{aligned}$$

The *equivalent lowpass signal* $x_{\text{LP}}(t)$ for the bandpass signal $x(t)$ is closely related to the complex envelope $\tilde{x}(t)$. We shall follow the lead of most texts which consider $x_{\text{LP}}(t)$ and $\tilde{x}(t)$ to be identical. However, it should be noted that some texts such as Ref. 9 define $x_{\text{LP}}(t)$ such that it equals $\tilde{x}(t)/2$. The usefulness of representing a signal in terms of its complex envelope will be discussed in Sec. 7.8.

Preenvelope [Refs. 5, 6, and 7]

If a signal $\underline{x}(t)$ has a Fourier transform $\underline{X}(f)$ which vanishes for negative frequencies, the signal will be complex-valued, and the real and imaginary parts will be related via the Hilbert transform:

$$\underline{x}(t) = x(t) + j\hat{x}(t) \qquad (6.83)$$

If $\underline{x}(t)$ and $x(t)$ are related in this manner, then the complex signal $\underline{x}(t)$ is called the *preenvelope*, or *analytic signal*, of the real signal $x(t)$. The Fourier transform $\underline{X}(f)$ of the analytic signal $\underline{x}(t)$ is related to the Fourier transform $X(f)$ of the real signal $x(t)$ via

$$\underline{X}(f) = \begin{cases} 2X(f) & f > 0 \\ X(f) & f = 0 \\ 0 & f < 0 \end{cases} \qquad (6.84)$$

The concept of preenvelope was first presented by Dugunji [Ref. 6] as an improvement over the "envelope" presented by Rice [Ref. 7]. For a bandpass signal $x(t)$, the complex envelope and preenvelope are related via

$$\underline{x}(t) = \tilde{x}(t) \exp(j\omega_0 t) \qquad (6.85)$$

The magnitudes of the complex envelope and preenvelope are each equal to the envelope $a(t)$:

$$|\underline{x}(t)| = |\tilde{x}(t)| = a(t) \qquad (6.86)$$

6.9 REFERENCES

1. S. Haykin: *Communication Systems*, 2d ed., Wiley, New York, 1983.
2. E. O. Brigham: *The Fast Fourier Transform*, Prentice-Hall, Englewood Cliffs, N.J., 1974.
3. A. A. Roberts and C. T. Mullis: *Digital Signal Processing*, Addison-Wesley, Reading, Mass., 1987.
4. B. Rorabaugh: *Signal Processing Design Techniques*, TAB Professional and Reference Books, Blue Ridge Summit, Penn., 1986.
5. H. Urkowitz: *Signal Theory and Random Processes*, Artech House, Dedham, Mass., 1983.

6. J. Dugunji: "Envelopes and Pre-envelopes of Real Waveforms," *IRE Trans. Inf. Theory*, IT-4, 1958, pp. 53–57.
7. S. O. Rice: "Mathematical Analysis of Random Noise," *Bell System Technical Journal*, July 1944, pp. 282–332.
8. M. Kanefsky, *Communication Techniques for Digital and Analog Signals*, Harper & Row, New York, 1985.
9. A. B. Carlson: *Communication Systems*, 3rd ed., McGraw-Hill, New York, 1981.

Chapter 7

System Theory

7.1 Systems [Refs. 1 and 3]

In the study of communication theory, a *system* can be generally regarded as anything which accepts one or more input signals and operates upon them to produce one or more output signals. When signals are represented as mathematical functions, it is convenient to represent systems as *operators* which operate upon input functions to produce output functions. Two alternative notations for representing a system H with input x and output y are given in Eqs. (7.1) and (7.2). Note that x and y can each be scalar-valued or vector-valued.

$$y = H[x] \qquad (7.1)$$

$$y = H x \qquad (7.2)$$

This book will use the notation of Eq. (7.1) as this is less likely to be confused with multiplication of x by a value H.

A system H can be represented pictorially in a flow diagram as shown in Fig. 7.1. For vector-valued x and y, the individual components are sometimes explicitly shown as in Fig. 7.2a or lumped together as shown in Fig. 7.2b. Sometimes, in order to emphasize their vector nature, the input and output are drawn as in Fig. 7.2c.

In different presentations of system theory, the notational schemes used exhibit some variation. The more precise treatments (such as Ref. 7) use x or $x(\cdot)$ to denote a function of time defined over the interval $(-\infty, \infty)$. A function defined over a more restricted interval such as $[t_0, t_1)$ would be denoted as $x_{[t0, t1)}$. The notation $x(t)$ is reserved for denoting the value of x at time t. Less precise treatments (such as Ref. 1) use $x(t)$ to denote both functions of time defined over $(-\infty, \infty)$

Figure 7.1 Pictorial representation of a system.

and the value of x at time t. When not evident from context, words of explanation must be included to indicate which particular meaning is intended. Using the less precise notational scheme, (7.1) could be rewritten as:

$$y(t) = H[x(t)] \qquad (7.3)$$

While it appears that the precise notation should be the more desirable, the relaxed conventions exemplified by (7.3) are widespread in the literature.

Linearity

If the relaxed system H is *homogeneous*, multiplying the input by a constant gain is equivalent to multiplying the output by the same constant gain, and the two configurations shown in Fig. 7.3 are

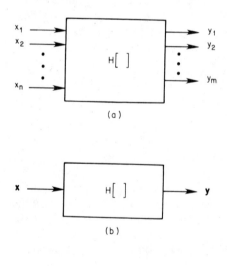

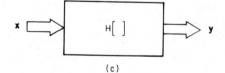

Figure 7.2 Pictorial representations of a system with multiple inputs and outputs.

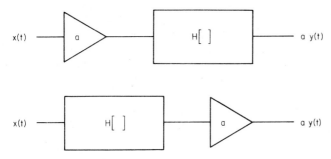

Figure 7.3 Homogeneous system.

equivalent. Mathematically stated, the relaxed system H is homogeneous if, for constant a,

$$H[ax] = a\, H[x] \tag{7.4}$$

If the relaxed system H is *additive*, the output produced for the sum of two input signals is equal to the sum of the outputs produced for each input individually, and the two configurations shown in Fig. 7.4 are equivalent. Mathematically stated, the relaxed system H is additive if

$$H[x_1 + x_2] = H[x_1] + H[x_2] \tag{7.5}$$

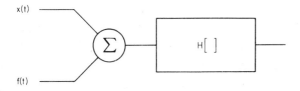

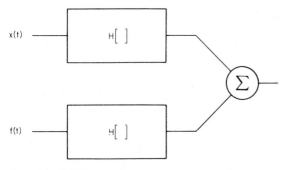

Figure 7.4 Additive system.

A system that is both homogeneous and additive is said to "exhibit *superposition*" or to "satisfy the principle of superposition." A system which exhibits superposition is called a *linear* system. Under certain restrictions, additivity implies homogeneity. Specifically the fact that a system H is additive implies that

$$H[\alpha x] = \alpha\, H[x] \tag{7.6}$$

for any rational α. Any real number can be approximated with arbitrary precision by a rational number; therefore, additivity implies homogeneity for real a provided that

$$\lim_{\alpha \to a} H[\alpha x] = H[ax] \tag{7.7}$$

Time invariance

A *time-invariant* system is a system whose characteristics do not change over time. A system is said to be relaxed if it is not still responding to any previously applied input. Given a relaxed system H such that

$$y(t) = H[x(t)] \tag{7.8}$$

then H is time-invariant if and only if

$$y(t - \tau) = H[x(t - \tau)] \tag{7.9}$$

for any τ and any $x(t)$. A time-invariant system is also called a *fixed*, or *stationary*, system. A system which is not time-invariant is called a *time-varying* system, *variable system*, or *nonstationary* system.

Causality

In a *causal* system, the output at time t can depend only upon the input at times t and prior. Mathematically stated, a system H is causal if and only if

$$H[x_1(t)] = H[x_2(t)] \quad \text{for } t \le t_0 \tag{7.10}$$

given that

$$x_1(t) = x_2(t) \quad \text{for } t \le t_0$$

A *noncausal*, or *anticipatory*, system is one in which the present output does depend upon future values of the input. Noncausal systems occur in theory, but they cannot exist in the real world. This is unfortunate, since we will often discover that some especially desirable frequency responses can be obtained only from noncausal systems. However, causal realizations can be created for noncausal systems in which the present output depends at most upon past, present, and a finite extent of future inputs. In such cases, a causal realization is obtained by simply delaying the output of the system for a finite interval until all the required inputs have entered the system and are available for determination of the output.

7.2 Characterization of Linear Systems
[Refs. 1 through 3]

A linear system can be characterized by a differential equation, step response, impulse response, complex-frequency domain system function, or a transfer function. The relationship between these various characterizations are given in Table 7.1.

Impulse response [Refs. 1, 2, 3, 5, and 6]

The *impulse response* of a system is the output response produced when a unit impulse $\delta(t)$ is applied to the input of the previously relaxed system. This is an especially convenient characterization of a linear system, since the response $y(t)$ to any continuous-time input

TABLE 7.1 Relationships between Characterizations of Linear Systems

Starting with	Perform	To obtain
Time domain differential equation relating $x(t)$ and $y(t)$	Laplace transform	Complex-frequency domain system function
	Compute $y(t)$ for $x(t)$ = unit impulse	Impulse response $h(t)$
	Compute $y(t)$ for $x(t)$ = unit step	Step response $a(t)$
Step response $a(t)$	Differentiate with respect to time	Impulse response $h(t)$
Impulse response $h(t)$	Integrate with respect to time	Step response $a(t)$
	Laplace transform	Transfer function $H(s)$
Complex-frequency domain system function	Solve for $H(s) = Y(s)/X(s)$	Transfer function $H(s)$
Transfer function $H(s)$	Inverse Laplace transform	Impulse response $h(t)$

signal $x(t)$ is given by

$$y(t) = \int_{-\infty}^{\infty} x(\tau) h(t, \tau) \, d\tau \qquad (7.11)$$

where $h(t, \tau)$ denotes the system's response at time t to an impulse applied at time τ. The integral in (7.11) is sometimes referred to as the *superposition integral*. The particular notation used indicates that, in general, the system is time varying. For a time-invariant system, the impulse response at time t depends only upon the time delay from τ to t; and we can redefine the impulse response to be a function of a single variable and denote it as $h(t - \tau)$. Equation (7.11) then becomes

$$y(t) = \int_{-\infty}^{\infty} x(\tau) h(t - \tau) \, d\tau \qquad (7.12)$$

Via the simple change of variables $\lambda = t - \tau$, Eq. (7.12) can be rewritten as:

$$y(t) = \int_{-\infty}^{\infty} x(t - \lambda) h(\lambda) \, d\lambda \qquad (7.13)$$

If we assume that the input is zero for $t < 0$, the lower limit of integration can be changed to zero; and if we further assume that the system is causal, the upper limit of integration can be changed to t, thus yielding

$$y(t) = \int_0^t x(\tau) h(t - \tau) \, d\tau = \int_0^t x(t - \lambda) h(\lambda) \, d\lambda \qquad (7.14)$$

The integrals in (7.14) are known as *convolution integrals*, and the equation indicates that "$y(t)$ equals the *convolution* of $x(t)$ and $h(t)$." It is often more compact and convenient to denote this relationship as:

$$y(t) = x(t) \otimes h(t) = h(t) \otimes x(t) \qquad (7.15)$$

Various texts use different symbols, such as stars or asterisks, in place of $\otimes$ to indicate convolution. The asterisk is probably favored by most printers, but in some contexts its usage to indicate convolution could be confused with the complex conjugation operator. A typical system's impulse response is sketched in Fig. 7.5.

Step response [Refs. 1 through 4]

The *step response* of a system is the output signal produced when a unit step $u(t)$ is applied to the input of the previously relaxed system.

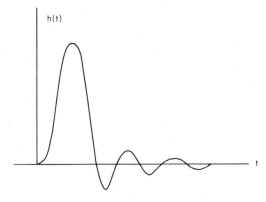

Figure 7.5 Impulse response of a typical system.

Since the unit step is simply the time integration of a unit impulse, it can easily be shown that the step response of a system can be obtained by integrating the impulse response. A typical system's step response is shown in Fig. 7.6.

7.3 Laplace Transform [Refs. 1, 2, 3, 4, and 10]

The *Laplace transform* is a technique which is useful for transforming differential equations into algebraic equations that can be more easily manipulated to obtain desired results. The Laplace transform is named for the French mathematician Pierre Simon de Laplace (1749–1827).

In most communications applications, the functions of interest will usually (but not always) be functions of time. The Laplace transform of a time function $x(t)$ is usually denoted as $X(s)$ or $\mathscr{L}[x(t)]$ and is

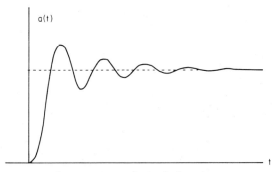

Figure 7.6 Step response of a typical system.

defined by

$$X(s) = \mathscr{L}[x(t)] = \int_0^\infty x(t)\, e^{-st}\, dt \qquad (7.16)$$

The complex variable s is usually referred to as *complex frequency* and is of the form $\sigma + j\omega$, where σ and ω are real variables sometimes referred to as *neper frequency* and *radian frequency*, respectively. The Laplace transform for a given function $x(t)$ is obtained by simply evaluating the given integral. Some mathematics texts such as Ref. 4 denote the time function with an uppercase letter and the frequency function with a lowercase letter. However, the use of lowercase for time functions is almost universal within the engineering literature.

If we transform both sides of a differential equation in t using the definition (7.16), we obtain an algebraic equation in s which can be solved for the desired quantity. The solved algebraic equation can then be transformed back into the time domain by using the inverse Laplace transform. The inverse Laplace transform is defined by

$$x(t) = \mathscr{L}^{-1}[X(s)] = \frac{1}{2\pi j}\int_C X(s)\, e^{st}\, ds \qquad (7.17)$$

where C is a contour of integration chosen so as to include all singularities of $X(s)$. The inverse Laplace transform for a given function $X(s)$ can be obtained by evaluating the given integral. However, this integration is often a major chore—when tractable, it will usually involve application of the residue theorem from the theory of complex variables. Fortunately, in most cases of practical interest, direct evaluation of (7.16) and (7.17) can be avoided by using some well-known transform pairs listed in Table 7.2 along with a number of transform properties presented in Sec. 7.4.

Example Find the Laplace transform of $x(t) = e^{-\alpha t}$.

solution

$$X(s) = \int_0^\infty e^{-\alpha t} e^{-st}\, dt \qquad (7.18)$$

$$= \int_0^\infty e^{-(\alpha + s)t}\, dt \qquad (7.19)$$

Using entry 1 from Table 2.8, we substitute $a = \alpha + s$ and obtain as a solution:

$$X(s) = \frac{1}{s + \alpha} \qquad (7.20)$$

Notice that this result agrees with entry 8 in Table 7.2.

TABLE 7.2 Laplace Transform Pairs

Ref. no.	$x(t)$	$X(s)$
1.	1	$\dfrac{1}{s}$
2.	$u_1(t)$	$\dfrac{1}{s}$
3.	$\delta(t)$	1
4.	t	$\dfrac{1}{s^2}$
5.	t^n	$\dfrac{n!}{s^{n+1}}$
6.	$\sin \omega t$	$\dfrac{\omega}{s^2+\omega^2}$
7.	$\cos \omega t$	$\dfrac{s}{s^2+\omega^2}$
8.	e^{-at}	$\dfrac{1}{s+a}$
9.	$e^{-at}\sin \omega t$	$\dfrac{\omega}{(s+a)^2+\omega^2}$
10.	$e^{-at}\cos \omega t$	$\dfrac{s+a}{(s+a)^2+\omega^2}$

7.4 Properties of the Laplace Transform
[Refs. 1 and 2]

Some properties of the Laplace transform are listed in Table 7.3. These properties can be used in conjunction with the transform pairs presented in Table 7.2 to obtain most of the Laplace transforms that will ever be needed in practical engineering situations. Some of the entries in the table require further explanation, which is provided below.

Time shifting right

Consider the function $f(t)$ shown in Fig. 7.7a. The function has nonzero values for $t < 0$, but since the one-sided Laplace transform integrates only over positive time, these values for $t < 0$ have no impact on the evaluation of the transform. If we now shift $f(t)$ to the right by τ units as shown in Fig. 7.7b, some of the nonzero values from the left of the origin will be moved to the right of the origin, where they will be

TABLE 7.3 Properties of the Laplace Transform

Property	Time function	Transform
1. Homogeneity	$af(t)$	$aF(s)$
2. Additivity	$f(t) + g(t)$	$F(s) + G(s)$
3. Linearity	$af(t) + bg(t)$	$aF(s) + bG(s)$
4. First derivative	$\dfrac{d}{dt}f(t)$	$sF(s) - f(0)$
5. Second derivative	$\dfrac{d^2}{dt^2}f(t)$	$sF(s) - sf(0) - \dfrac{d}{dt}f(0)$
6. kth derivative	$\dfrac{d^{(k)}}{dt^k}f(t)$	$s^k F(s) - \sum_{n=0}^{k-1} s^{k-1-n} f^{(n)}(0)$
7. Integration	$\int_{-\infty}^{t} f(\tau)\,d\tau$	$\dfrac{F(s)}{s} + \dfrac{1}{s}\left(\int_{-\infty}^{t} f(\tau)\,d\tau\right)_{t=0}$
	$\int_{0}^{t} f(\tau)\,d\tau$	$\dfrac{F(s)}{s}$
8. Frequency shift	$e^{-at}f(t)$	$X(s+a)$
9. Time shift right	$u_1(t-\tau)f(t-\tau)$	$e^{-\tau s}F(s) \quad a>0$
10. Time shift left	$f(t+\tau), f(t)=0 \text{ for } 0<t<\tau$	$e^{\tau s}F(s)$
11. Convolution	$y(t) = \int_0^t h(t-\tau)x(\tau)\,d\tau$	$Y(s) = H(s)X(s)$
12. Multiplication	$f(t)g(t)$	$\dfrac{1}{2\pi j}\int_{c-j\infty}^{c+j\infty} F(s-r)G(r)\,dr$
		$\sigma_g < c < \sigma - \sigma_f$

Notes: $f^{(k)}(t)$ denotes the kth derivative of $f(t)$.
$f^{(0)}(t) = f(t)$.

included in the evaluation of the transform. The Laplace transform's properties with regard to a time shift right must be stated in such a way that these previously unincluded values will not be included in the transform of the shifted function either. This can be easily accomplished through multiplying the shifted function $f(t - \tau)$ by a shifted unit step function $u_1(t - \tau)$ as shown in Fig. 7.7c. Thus we have

$$\mathscr{L}\{u_1(t-\tau)f(t-\tau)\} = e^{-\tau s}F(s) \qquad a>0 \qquad (7.21)$$

Time shift left

Consider now the case when $f(t)$ is shifted to the left. Such a shift will move a portion of $f(t)$ from positive time where it is included in the transform evaluation into negative time where it will not be included

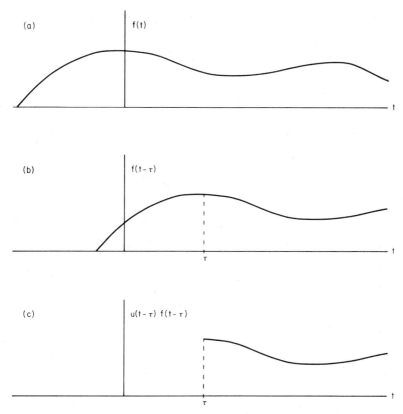

Figure 7.7 Signals for explanation of the Laplace transform's "time shift right" property.

in the transform evaluation. The Laplace transform's properties with regard to a time shift left must be stated in such a way that all included values from the unshifted function will likewise be included in the transform of the shifted function. This can be accomplished by requiring that the original function be equal to zero for all values of t from zero to τ, if a shift to the left by τ units is to be made. Thus for a shift left by τ units:

$$\mathscr{L}\{f(t+\tau)\} = F(s)\, e^{\tau s} \quad \text{if } f(t) = 0 \quad \text{for } 0 < t < \tau \quad (7.22)$$

7.5 Transfer Functions [Refs. 1, 2, and 3]

The *transfer function* H(s) of a system is equal to the Laplace transform of the output signal divided by the Laplace transform of the input signal:

$$H(s) = \frac{Y(s)}{X(s)} = \frac{\mathscr{L}[y(t)]}{\mathscr{L}[x(t)]} \qquad (7.23)$$

It can be shown that the transfer function is also equal to the Laplace transform of the system's impulse response:

$$H(s) = \mathscr{L}[h(t)] \qquad (7.24)$$

Therefore, $\qquad y(t) = \mathscr{L}^{-1}\{H(s)\mathscr{L}[x(t)]\} \qquad (7.25)$

Equation (7.25) presents an alternative to the convolution defined by Eq. (7.14) for obtaining a system's response $y(t)$ to any input $x(t)$, given the impulse response $h(t)$. Simply perform the following steps:

1. Compute $H(s)$ as the Laplace transform of $h(t)$.
2. Compute $X(s)$ as the Laplace transform of $x(t)$.
3. Compute $Y(s)$ as the product of $H(s)$ and $X(s)$.
4. Compute $y(t)$ as the inverse Laplace transform of $Y(s)$. (The Heaviside expansion presented in Sec. 7.6 is a convenient technique for performing the inverse transform operation.)

A transfer function defined as in (7.23) can be put into the form:

$$H(s) = \frac{P(s)}{Q(s)} \qquad (7.26)$$

where $P(s)$ and $Q(s)$ are polynomials in s. For $H(s)$ to be stable and realizable in the form of a lumped-parameter network, it can be shown [Ref. 2] that all the coefficients in the polynomials $P(s)$ and $Q(s)$ must be real. Furthermore, all the coefficients in $Q(s)$ must be positive. The polynomial $Q(s)$ must have a nonzero term for each degree of s from the highest to the lowest, unless all even-degree terms or all odd-degree terms are missing. If $H(s)$ is voltage ratio or current ratio (i.e., the input and output are either both voltages or both currents), the maximum degree of s in $P(s)$ cannot exceed the maximum degree of s in $Q(s)$. If $H(s)$ is a transfer impedance (i.e., the input is a current and the output is a voltage) or a transfer admittance (i.e., the input is a voltage and the output is a current), then the maximum degree of s in $P(s)$ can exceed the maximum degree of s in $Q(s)$ by at most 1. Note that these are only upper limits on the degree of s in $P(s)$; in either case, the maximum degree of s in $P(s)$ may be as small as zero. Also note that these are necessary, but not sufficient, conditions for H(s) to be a valid transfer function. A candidate H(s) satisfying all these conditions may still not be realizable as a lumped-parameter network.

Example Consider the following alleged transfer functions:

$$H_1(s) = \frac{s^2 - 2s + 1}{s^3 - 3s^2 + 3s + 1} \tag{7.27}$$

$$H_2(s) = \frac{s^4 + 2s^3 + 2s^2 - 3s + 1}{s^3 + 3s^2 + 3s + 2} \tag{7.28}$$

$$H_3(s) = \frac{s^2 - 2s + 1}{s^3 + 3s^2 + 1} \tag{7.29}$$

Equation (7.27) is not acceptable because the coefficient of s^2 in the denominator is negative. If Eq. (7.28) is intended as a voltage or current transfer ratio, it is not acceptable because the degree of the numerator exceeds the degree of the denominator. However, if Eq. (7.28) represents a transfer impedance or transfer admittance, it may be valid since the degree of the numerator exceeds the degree of the denominator by just 1. Equation (7.29) is not acceptable because the term for s is missing from the denominator.

A system's transfer function can be manipulated to provide a number of useful characterizations of the system's behavior. These characterizations are listed in Table 7.4 and examined in more detail in subsequent sections.

Some authors, such as Van Valkenburg [Ref. 2], use the term *network function* in place of *transfer function*.

TABLE 7.4 System Characterizations Obtained from the Transfer Function

Starting with	Perform	To obtain
Transfer function H(s)	Compute roots of H(s) denominator	Pole locations
	Compute roots of H(s) numerator	Zero locations
	Compute $\|H(j\omega)\|$ over all ω	Magnitude response $A(\omega)$
	Compute $\arg\{H(j\omega)\}$ over all ω	Phase response $\theta(\omega)$
Phase response $\theta(\omega)$	Divide by ω	Phase delay $\tau_p(\omega)$
	Differentiate with respect to ω	Group delay $\tau_g(\omega)$

7.6 Heaviside Expansion

General case

The Heaviside expansion provides a straightforward computational method for obtaining the inverse Laplace transform of certain types of complex-frequency functions. The function to be inverse-transformed must be expressed as a ratio of polynomials in s, where the order of the denominator polynomial exceeds the order of the numerator polynomial. If

$$H(s) = K_0 \frac{P(s)}{Q(s)} \qquad (7.30)$$

where $Q(s) = \prod_{k=1}^{n} (s - s_k)^{m_k} = (s - s_1)^{m_1}(s - s_2)^{m_2} \cdots (s - s_n)^{m_n} \qquad (7.31)$

then inverse transformations via the Heaviside expansion yields

$$\mathscr{L}^{-1}[H(s)] = K_0 \sum_{r=1}^{n} \sum_{k=1}^{m_r} [K_{rk} t^{m_r - k} \exp(s_r t)] \qquad (7.32)$$

where $K_{rk} = \dfrac{1}{(k-1)!(m_r - k)!} \dfrac{d^{k-1}}{ds^{k-1}} \left[\dfrac{(s - s_r)^{m_r} P(s)}{Q(s)} \right]_{s = s_r} \qquad (7.33)$

A method for computing the derivative in (7.33) can be found in Sec. 2.4.

Simple pole case

The complexity of the expansion given in (7.32) is significantly reduced for the case of $Q(s)$ having no repeated roots. The denominator of (7.30) is then given by

$$Q(s) = \prod_{k=1}^{n} (s - s_k) = (s - s_1)(s - s_2) \cdots (s - s_n) \qquad s_1 \neq s_2 \neq s_3 \neq \cdots \neq s_n$$
$$(7.34)$$

Inverse transformation via the Heaviside expansion then yields

$$\mathscr{L}^{-1}[H(s)] = K_0 \sum_{r=1}^{n} K_r e^{s_r t} \qquad (7.35)$$

where $K_r = \left[\dfrac{(s - s_r)P(s)}{Q(s)} \right]_{s = s_r} \qquad (7.36)$

The Heaviside expansion is named for Oliver Heaviside (1850–1925), an English physicist and electrical engineer who was the nephew of Charles Wheatstone (as in Wheatstone bridge).

7.7 Poles and Zeros [Refs. 1 through 4 and 7 through 9]

As pointed out previously, the transfer function for a realizable linear time-invariant system can always be expressed as a ratio of polynomials in s:

$$H(s) = \frac{P(s)}{Q(s)} \tag{7.37}$$

The numerator and denominator can each be factored to yield

$$H(s) = H_0 \frac{(s-z_1)(s-z_2)(s-z_3)\cdots(s-z_m)}{(s-p_1)(s-p_2)(s-p_3)\cdots(s-p_n)} \tag{7.38}$$

where the roots $z_1, z_2, \ldots, z_m$ of the numerator are called *zeros* of the transfer function, and the roots $p_1, p_2, \ldots, p_n$ of the denominator are called *poles* of the transfer function. Together, poles and zeros can be collectively referred to as *critical frequencies*. Each factor $(s - z_i)$ is called a *zero factor*, and each factor $(s - p_i)$ is called a *pole factor*. A repeated zero appearing n times is called either an *nth-order zero* or a *zero of multiplicity n*. Likewise, a repeated pole appearing n times is called either an *nth-order pole* or a *pole of multiplicity n*. Nonrepeated poles or zeros are sometimes described as *simple* or *distinct* to emphasize their nonrepeated nature.

Example Consider the transfer function given by

$$H(s) = \frac{s^3 + 5s^2 + 8s + 4}{s^3 + 13s^2 + 59s + 87} \tag{7.39}$$

The numerator and denominator can be factored to yield

$$H(s) = \frac{(s+2)^2(s+1)}{(s+5+2j)(s+5-2j)(s+3)} \tag{7.40}$$

Examination of (7.40) reveals that

$s = -1$ is a simple zero

$s = -2$ is a second-order zero

$s = -5 + 2j$ is a simple pole

$s = -5 - 2j$ is a simple pole

$s = -3$ is a simple pole

A system's poles and zeros can be depicted graphically as locations in a complex plane as shown in Fig. 7.8. In mathematics, the complex plane itself is called the *gaussian plane*, while a plot depicting

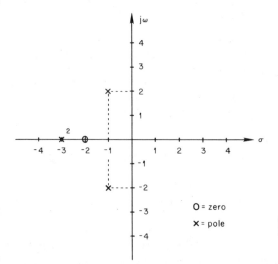

Figure 7.8 Plot of pole and zero locations.

complex values as points in the plane is called an *Argand diagram* or a *Wessel-Argand-Gaussian diagram*. In the 1798 transactions of the Danish academy, Caspar Wessel (1745–1818) published a technique for graphical representation of complex numbers, and Jean Robert Argand published a similar technique in 1806. Geometric interpretation of complex numbers played a central role in the doctoral thesis of Gauss.

Pole locations can provide convenient indications of a system's behavior as indicated in Table 7.5. Furthermore, poles and zeros

TABLE 7.5 Impact of Pole Locations upon System Behavior

Pole	Corresponding natural response component	Corresponding description of system behavior
Single real, negative	Decaying exponential	Stable
Single real, positive	Divergent exponential	Divergent instability
Real pair, negative, unequal	Decaying exponential	Overdamped (stable)
Real pair, negative, equal	Decaying exponential	Critically damped (stable)
Complex conjugate pair with negative real parts	Exponentially decaying sinusoid	Underdamped (stable)
Complex conjugate pair with zero real parts	Sinusoid	Undamped (marginally stable)
Complex conjugate pair with positive real parts	Exponentially saturating sinusoid	Oscillatory instability

possess the following properties which can sometimes be used to expedite the analysis of a system:

1. For real H(s), complex or imaginary poles and zeros will each occur in complex conjugate pairs which are symmetric about the σ axis.
2. For H(s) having even symmetry, the poles and zeros will exhibit symmetry about the $j\omega$ axis.
3. For nonnegative H(s), any zeros on the $j\omega$ axis will occur in pairs.

7.8 Magnitude, Phase, and Delay Responses
[Refs. 3 and 10]

A system's *steady-state response* $H(j\omega)$ can be determined by evaluating the transfer function H(s) at $s = j\omega$:

$$H(j\omega) = |H(j\omega)| e^{j\theta(\omega)} = H(s)|_{s=j\omega} \qquad (7.41)$$

The *magnitude response* is simply the magnitude of $H(j\omega)$:

$$|H(j\omega)| = (\{Re[H(j\omega)]\}^2 + \{Im[H(j\omega)]\}^2)^{1/2} \qquad (7.42)$$

It can be shown that

$$|H(j\omega)|^2 = H(s)H(-s)|_{s=j\omega} \qquad (7.43)$$

If H(s) is available in factored form as given by

$$H(s) = H_0 \frac{(s-z_1)(s-z_2)(s-z_3)\cdots(s-z_m)}{(s-p_1)(s-p_2)(s-p_3)\cdots(s-p_n)} \qquad (7.44)$$

then the magnitude response can be obtained by replacing each factor with its absolute value evaluated at $s = j\omega$:

$$|H(j\omega)| = H_0 \frac{|j\omega - z_1| \cdot |j\omega - z_2| \cdot |j\omega - z_3| \cdots |j\omega - z_m|}{|j\omega - p_1| \cdot |j\omega - p_2| \cdot |j\omega - p_3| \cdots |j\omega - p_n|} \qquad (7.45)$$

The *phase response* $\theta(\omega)$ is given by

$$\theta(\omega) = \tan^{-1}\left\{\frac{Im[H(j\omega)]}{Re[H(j\omega)]}\right\} \qquad (7.46)$$

Phase delay [Refs. 3 and 10]

The *phase delay* $\tau_p(\omega)$ of a system is defined as:

$$\tau_p(\omega) = \frac{-\theta(\omega)}{\omega} \qquad (7.47)$$

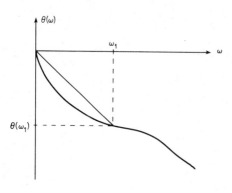

Figure 7.9 Phase delay.

where $\theta(\omega)$ is the phase response defined in Eq. (7.46). When evaluated at any specific frequency ω_1, Eq. (7.47) will yield the time delay experienced by a sinusoid of frequency ω passing through the system. Some authors define $\tau_p(\omega)$ without the minus sign shown on the RHS of (7.47). As illustrated in Fig. 7.9, the phase delay at a frequency ω_1 is equal to the negative slope of a secant drawn from the origin to the phase response curve at ω_1.

Group delay [Refs. 3 and 10]

The *group delay* $\tau_g(\omega)$ of a system is defined as:

$$\tau_g(\omega) = -\frac{d}{dt}\theta(\omega) \tag{7.48}$$

where $\theta(\omega)$ is the phase response defined in (7.46). In the case of a modulated carrier passing through the system, the modulation envelope will be delayed by an amount which is, in general, not equal to the delay $\tau_p(\omega)$ that is experienced by the carrier. If the system exhibits constant group delay over the entire bandwidth of the modulated signal, then the envelope will be delayed by an amount equal to τ_g. If the group delay is not constant over the entire bandwidth of the signal, the envelope will be distorted. As shown in Fig. 7.10, the group delay at a frequency ω_1 is equal to the negative slope of a tangent to the phase response at ω_1.

Assuming that the phase response of a system is sufficiently smooth, it can be approximated as:

$$\theta(\omega + \omega_c) = \tau_p \omega_c + \tau_g \omega_c \tag{7.49}$$

If an input signal $x(t) = a(t) \cos \omega_c t$ is applied to a system for which (7.49) holds, the output response will be given by

$$y(t) = Ka(t - \tau_g) \cos[\omega_c(t - \tau_p)] \tag{7.50}$$

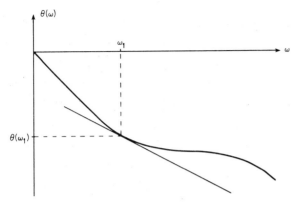

Figure 7.10 Group delay.

Since the envelope $a(t)$ is delayed by τ_g, the group delay is also called *envelope delay*. Likewise, since the carrier is delayed by τ_p, the phase delay is also called *carrier delay*.

7.9 Bandpass Systems

For a linear bandpass system having an impulse response of $h(t)$, there exists a corresponding lowpass signal with impulse response $\tilde{h}(t)$ such that

$$h(t) = 2Re[\tilde{h}(t) \exp(j\omega_c t)] \quad (7.51)$$

The impulse response $\tilde{h}(t)$ can be obtained as the inverse Fourier transform of $\tilde{H}(f)$, where $\tilde{H}(f)$ is defined as:

$$\tilde{H}(f - f_c) = \begin{cases} H(f) & f > 0 \\ 0 & f < 0 \end{cases} \quad (7.52)$$

A bandpass system's output response due to a bandpass input $x(t)$ is given by the convolution integral:

$$y(t) = \int_{-\infty}^{\infty} x(t) h(t - \tau) \, d\tau \quad (7.53)$$

where $y(t) = Re[\tilde{y}(t) \exp(\omega_c t)]$

$x(t) = Re[\tilde{x}(t) \exp(\omega_c t)]$

$h(t) = Re[\tilde{h}(t) \exp(\omega_c t)]$

Given (7.53), it can be shown [Ref. 11] that

$$\tilde{y}(t) = \int_{-\infty}^{\infty} \tilde{x}(t)\tilde{h}(t-\tau)\,d\tau \qquad (7.54)$$

Thus it is possible to analyze bandpass signals and systems entirely in terms of their lowpass equivalents. This fact will be exploited throughout the remainder of this book.

7.10 REFERENCES

1. R. J. Schwartz and B. Friedland: *Linear Systems*, McGraw-Hill, New York, 1965.
2. M. E. Van Valkenburg: *Network Analysis*, Prentice-Hall, Englewood Cliffs, N.J., 1974.
3. B. Rorabaugh: *Signal Processing Design Techniques*, TAB Professional and Reference Books, Blue Ridge Summit, Penn., 1986.
4. M. R. Spiegel: *Laplace Transforms*, Schaum's Outline Series, McGraw-Hill, New York, 1965.
5. S. Haykin: *Communication Systems*, 2d ed., Wiley, New York, 1983.
6. G. R. Cooper and C. D. McGillem: *Modern Communications and Spread Spectrum*, McGraw-Hill, New York, 1986.
7. C-T. Chen: *Linear System Theory and Design*, Holt, Rinehart and Winston, New York, 1984.
8. H. L. Van Trees: *Detection, Estimation, and Modulation Theory—Part I: Detection, Estimation, and Linear Modulation Theory*, Wiley, New York, 1968.
9. C. B. Boyer: *A History of Mathematics*, Wiley, New York, 1968.
10. H. J. Blinchikoff and A. I. Zverev: *Filtering in the Time and Frequency Domains*, Wiley, New York, 1976.
11. J. G. Proakis: *Digital Communications*, McGraw-Hill, New York, 1983.

Chapter 8

Noise

8.1 White Noise

White noise is an idealized noise process having a power spectral density that is constant over all frequencies as shown in Fig. 8.1. The term *white* derives from the fact that the spectrum of white light is constant over all frequencies in the visible range. By convention, the constant value of the two-sided psd for white noise is usually denoted as $N_0/2$. The factor of 2 in the denominator is a convenience so that the power passed by an ideal lowpass filter (LPF) having a bandwidth of B will be equal to $N_0 B$. In a similar vein, the constant value of the one-sided psd for white noise is denoted as N_0. When the noise generated within a system is modeled as white noise applied to the system's input, the value of N_0 is simply kT where k is Boltzmann's constant and T is the equivalent noise temperature of the system.

As discussed in Sec. 5.4, the power spectral density and autocorrelation function form a Fourier transform pair. Using pair 3 from Table 6.3, we conclude that the acf of white noise is an impulse or delta function located at the origin.

$$\frac{N_0}{2}\delta(\tau) \overset{\text{FT}}{\leftrightarrow} \frac{N_0}{2} \qquad (8.1)$$

Physically, an impulse at the origin signifies that white noise has infinite average power. Obviously, no noise process in the real world can have infinite average power, which is why the word *idealized* appears above in the opening sentence. Nevertheless, the concept of white noise is a mathematical convenience that finds widespread use in theoretical work. Once white noise is (conceptually) passed through a filter of finite bandwidth, the objectionable attribute of

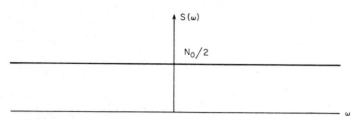

Figure 8.1 Power spectral density of white noise.

infinite average power disappears. Although it theoretically need not be so, whenever a white noise process is assumed, it is almost always assumed to also be a gaussian process and is called *white gaussian noise* (WGN) or *additive white gaussian noise* (AWGN).

Example Ideal lowpass filtering of white noise is a classic example which is presented in numerous texts [Refs. 6, 7, 12, 13, and 14]. The transfer function of the ideal LPF is given by

$$H(f) = \begin{cases} 1 & |f| \leq B \\ 0 & \text{elsewhere} \end{cases}$$

If white noise with a psd of $N_0/2$ is applied to the input of such a filter, the psd of the output noise process will be given by

$$S(f) = \begin{cases} \dfrac{N_0}{2} & |f| \leq B \\ 0 & \text{elsewhere} \end{cases}$$

Using pair 12 from Table 6.3, we find the autocorrelation of the output noise is

$$R(\tau) = N_0 B \operatorname{sinc}(2B\tau)$$

The average power is equal to $R(\tau)$ evaluated at $\tau = 0$ or simply $N_0 B$—an intuitively pleasing result. The output noise process is referred to as *band-limited white noise*. In order to avoid the verbal tap dancing needed to deal with infinite average power, it may be helpful to consider band-limited white noise first and then simply think of ideal white noise as a limiting process that is approached as the bandwidth approaches infinity.

8.2 Noise Equivalent Bandwidth

The magnitude response $|H(f)|$ of an arbitrary lowpass filter is sketched (solid line) in Fig. 8.2. If zero-mean white noise having a (two-sided) psd of $N_0/2$ is applied to the input of such a filter, the resulting output will have a finite average power that is given by

$$N = \frac{N_0}{2} \int_{-\infty}^{\infty} |H(f)|^2 \, df \tag{8.2}$$

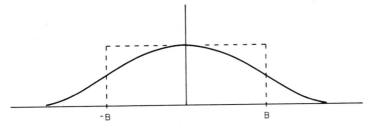

Figure 8.2 Illustration of noise equivalent bandwidth.

Figure 8.2 also shows (dashed line) the magnitude response of an ideal lowpass filter. The *noise equivalent bandwidth* of the arbitrary filter is defined as the value of B for which the ideal filter and arbitrary filter produce the same output power from identical white noise inputs. The average output noise power from the ideal filter is

$$N = N_0 B H^2(0) \tag{8.3}$$

The value of B for which the two filters produced the same output noise power is easily found by equating (8.2) and (8.3) and solving for B to yield

$$B = \frac{\int_{-\infty}^{\infty} |H(f)|^2 \, df}{2H^2(0)} = \frac{\int_{0}^{\infty} |H(f)|^2 \, df}{H^2(0)} \tag{8.4}$$

8.3 Thermal Noise

The random motion of electrons in a conductor causes an electrical noise called *thermal noise* which is gaussian-distributed with zero mean. The variance is given by

$$\sigma^2 = E[v^2] = \frac{2R(\pi k T)^2}{3h} \quad V^2 \tag{8.5}$$

where $k =$ Boltzmann's constant $\approx 1.380622 \times 10^{-23}$ J/K (joules per kelvin)
$h =$ Planck's constant $\approx 6.626196 \times 10^{-34}$ J·s (joule·second)
$T =$ temperature in kelvins
$R =$ resistance in ohms

(*Note*: The official SI unit of absolute temperature is simply "kelvin," even though "degrees kelvin" sounds more natural.)

The spectral density of the mean-square noise voltage is given by

$$S(f) = \frac{2Rh|f|}{\exp[h|f|/(kT)] - 1} \quad \text{V}^2/\text{Hz} \tag{8.6}$$

For $|f| \ll kT/h$, Eq. (8.6) can be approximated as:

$$S(f) \approx 2RkT\left(1 - \frac{h|f|}{2kT}\right) \tag{8.7}$$

For frequencies up to approximately 10^{12} Hz, this expression is nearly constant and can be further simplified to

$$S(f) = 2RkT \tag{8.8}$$

In a bandwidth of B Hz, the mean-square value of the noise voltage is given by

$$E[v^2] = 2BS(f) = 4kTRB \tag{8.9}$$

A noisy resistor can be modeled as an ideal resistor in series with a noise voltage source having a mean-square voltage as given by (8.9). The *available power* of a source is defined as the power delivered to a load which is matched to the impedance of the source. Thus the available power will be the power delivered by the resistor and voltage source combination to a load resistance R. The load current will be $v/(2R)$, and the voltage drop across the load will be $v/2$. Thus the available power is equal to $v^2/(4R)$. Using the expected value of v^2 given by (8.1), we find that the available power is simply kTB watts.

Equivalent noise temperature

All actual systems generate noise internally. *Equivalent noise temperature* is one way of characterizing the amount of noise generated in a linear two-port system. Imagine that it is possible to obtain a noiseless version of the system under consideration. Conceptually, we could then connect a resistor to the input and then adjust the temperature of the resistor until the thermal noise generated by the resistor causes the available noise power at the output of the idealized system to exactly equal the noise power at the output of the actual system. The temperature of the resistor then defines the equivalent noise temperature of the system. Note that this definition assumes that the resistor is matched to the input impedance of the system.

Noise figure

The *spot noise figure* F of a linear two-port system at an operating frequency of f is defined as:

$$F = \frac{S_{\text{out}}(f)}{G(f)S_{\text{in}}(f)} = \frac{G(f)S_{\text{in}}(f) + S_g(f)}{G(f)S_{\text{in}}(f)} \quad (8.10)$$

where $G(f)$ = available power gain of the system
   $S_{\text{in}}(f)$ = power spectral density of the input noise
   $S_{\text{out}}(f)$ = power spectral density of the noise at the output
   $S_g(f)$ = psd of the portion of output noise that is generated internally within the system

The *average noise figure* F_{avg} is defined as:

$$F_{\text{avg}} = \frac{\int_{-\infty}^{\infty} S_{\text{out}}(f)\, df}{\int_{-\infty}^{\infty} G(f)S_{\text{in}}(f)\, df} \quad (8.11)$$

If we make three assumptions:

1. The noise applied to the input is modeled as thermal noise generated by a resistor which is matched to the system input and which is at a temperature of T
2. The system has constant gain G across a passband of width B and a gain of zero outside of this passband
3. The equivalent noise temperature of the system is T_e

then Eq. (8.11) can be rewritten as:

$$F_{\text{avg}} = \frac{T + T_e}{T}$$

8.4 Quadrature Form Representation of Bandpass Noise [Refs. 6 and 7]

General case

Consider a bandpass noise process $n(t)$ which has a power spectral density as depicted in Fig. 8.3. Such a process can be expressed in quadrature form relative to a carrier of frequency f_c and phase ψ:

$$n(t) = n_c(t) \cos(2\pi f_c t) - n_s(t) \sin(2\pi f_c t) \quad (8.12)$$

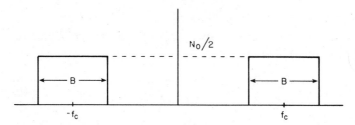

Figure 8.3 Spectrum of bandpass noise produced by ideal filtering of white noise.

The *envelope* of $n(t)$ is given by

$$A(t) = \sqrt{n_c^2(t) + n_s^2(t)} \qquad (8.13)$$

The *instantaneous phase* of $n(t)$ is given by

$$\theta(t) = \tan^{-1}\left(\frac{n_s(t)}{n_c(t)}\right) \qquad (8.14)$$

It can be shown that the quadrature form representation of Eq. (8.12) exhibits the following properties:

1. If $n(t)$ is a bandpass process of bandwidth $2B$ centered at f_c, the *inphase* component $n_c(t)$ and *quadrature* component $n_s(t)$ are each lowpass processes of (one-sided) bandwidth B. Specifically, if $n(t)$ has a power spectral density $S_n(f)$ which occupies frequencies in the interval $(f_c - B) \leq |f| \leq (f_c + B)$, then the power spectral densities $S_c(f)$ and $S_s(f)$ of the inphase and quadrature component, respectively, are given by

$$S_c(f) = S_s(f) = \begin{cases} S_n(f+f_c) + S_n(f-f_c) & |f| \leq B \\ 0 & \text{elsewhere} \end{cases} \qquad (8.15)$$

2. The two components $n_c(t)$ and $n_s(t)$ are uncorrelated.
3. If $n(t)$ has zero mean, then $n_c(t)$ and $n_s(t)$ each have zero mean.
4. If $n(t)$ has zero mean, then the variances of $n_c(t)$ and $n_s(t)$ will each equal the variance of $n(t)$.

$$\sigma_n^2 = \sigma_c^2 = \sigma_s^2$$

5. If $n(t)$ is wide-sense stationary, then $n_c(t)$ and $n_s(t)$ are jointly wide-sense stationary.

Gaussian case [Refs. 2 and 6]

If $n(t)$ is a bandpass gaussian process, then the following properties will hold in addition to the properties given above.

1. The inphase component $n_c(t)$ and the quadrature component $n_s(t)$ will be jointly gaussian, as well as statistically independent.
2. The envelope $A(t)$ will be a random process in which the ensemble random variables are Rayleigh distributed.
3. The phase $\theta(t)$ will be a random process in which the ensemble random variables are uniformly distributed from zero to 2π.

8.5 Superposition of Noise Powers [Ref. 6]

The power in a noise process $n(t)$ is given by the variance of the process:

$$P = E[n^2(t)]$$

If two noise processes $n_1(t)$ and $n_2(t)$ are added together, the power in the resulting process is given by

$$P = E\{[n_1(t) + n_2(t)]^2\}$$
$$= E[n_1^2(t)] + 2E[n_1(t)n_2(t)] + E[n_2^2(t)]$$

If $n_1(t)$ and $n_2(t)$ are uncorrelated, then $E[n_1(t)n_2(t)] = 0$, and the power of the summed processes equals the sum of the individual powers of the processes.

$$P = E[n_1^2(t)] + E[n_2^2(t)] = P_1 + P_2$$

In such cases it can be said that *superposition of power applies* or that P_1 and P_2 *add on a power basis*.

8.6 Amplitude of a Sine Wave with Random Phase [Refs. 1 and 9]

Consider a sinusoid of the form:

$$x = A \sin(\omega t + \theta) \tag{8.16}$$

where A is constant and θ is uniformly distributed on $(0, 2\pi)$. The instantaneous amplitude x will have a probability density function

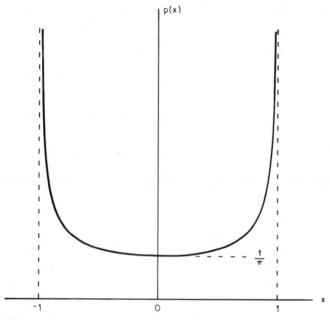

Figure 8.4 Probability density function for the amplitude of a sine wave with random phase.

given by

$$p(x) = \left(\frac{1}{\pi}\right)(A^2 - x^2)^{-1/2} \qquad (8.17)$$

Notice that (8.17) does not depend on ω or t. A plot of $p(x)$ is shown in Fig. 8.4.

The characteristic function $\phi_y(\omega)$ of y is given by

$$\phi_y(\omega) = \frac{1}{2\pi} \int_0^{2\pi} \exp[j\omega A \sin(\omega t + \theta)] \, d\theta$$
$$= J_0(A\omega) \qquad (8.18)$$

where J_0 denotes the zero-order Bessel function described in Sec. 2.7.

8.7 Amplitude of a Sine Wave Plus Gaussian Noise [Refs. 1 through 5]

Consider a signal:

$$x(t) = A \sin(\omega t + \theta) + n(t) \qquad (8.19)$$

where $n(t)$ = zero-mean gaussian noise process with variance σ^2
θ = uniformly distributed on $(0, 2\pi)$
$n(t)$ and θ are statistically independent

The instantaneous amplitude x will have a probability density function given by

$$p(x) = \frac{1}{\sigma\sqrt{2\pi}} \sum_{n=0}^{\infty} \left\{ \frac{[-x^2/(2\sigma^2)]^n}{n!} {}_1F_1\left(n + \frac{1}{2}; 1; \frac{-A^2}{2\sigma^2}\right) \right\} \quad (8.20)$$

where ${}_1F_1(\cdot)$ denotes the confluent hypergeometric function discussed in Sec. 2.8. The pdf given by (8.20) is often simplified by defining $z = x/\sigma$ and $a = A/\sigma$. Then (8.20) can be rewritten for the normalized variable z as:

$$p(z) = \frac{1}{\sqrt{2\pi}} \sum_{n=0}^{\infty} \left[\frac{(-z^2)^n}{2^n n!} {}_1F_1\left(n + \frac{1}{2}; 1; \frac{-a^2}{2}\right) \right] \quad (8.21)$$

The power in the sine wave equals $A^2/2$, and the noise power equals σ^2; therefore, the signal-to-noise ratio (SNR) equals $a^2/2$. Plots of (8.21) for various values of the SNR are shown in Fig. 8.5. The

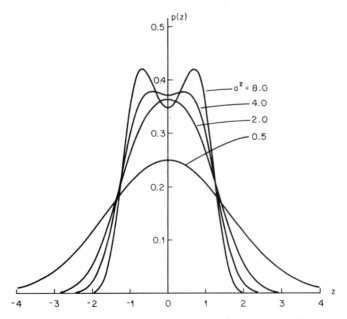

Figure 8.5 Probability density functions for the amplitude of a sine wave plus gaussian noise.

characteristic function of x is given by

$$\phi_x(\omega) = J_0(A\omega) \exp\left(\frac{-\sigma^2\omega^2}{2}\right) \quad (8.22)$$

where J_0 denotes the zero-order Bessel function discussed in Sec. 2.7.

Background. At any given time instant t_i, the amplitude $x(t_i)$ equals the sum of the instantaneous amplitude of a randomly phased sinusoid plus the random variable which represents an instantaneous sample from a gaussian noise process. Since these two components are statistically independent, the characteristic function of their sum will be given by the product of their individual characteristic functions.

$$\phi_s(\omega) = J_0(A\omega)$$

$$\phi_n(\omega) = \exp\left(\frac{-\sigma^2\omega^2}{2}\right)$$

$$\phi_x(\omega) = \phi_s(\omega)\phi_n(\omega) = J_0(A\omega) \exp\left(\frac{-\sigma^2\omega^2}{2}\right) \quad (8.23)$$

It can be easily verified that (8.23) equals the inverse Fourier transform of (8.20).

8.8 Envelope of a Sine Wave Plus Bandpass Gaussian Noise [Refs. 1, 2, 4, 5, and 8]

Consider a signal:

$$x(t) = A\cos(\omega t + \theta) + n(t) \quad (8.24)$$

where $n(t)$ is a zero-mean bandpass gaussian noise process. It is possible to rewrite (8.24) in quadrature form as:

$$x(t) = [A\cos(\theta) + n_I(t)]\cos(\omega t) - [A\sin(\theta) + n_Q(t)]\sin(\omega t) \quad (8.25)$$

Comparing (8.25) to Eq. (6.76), we can see that the envelope is given by

$$v(t) = \{[A\cos(\theta) + n_I(t)]^2 + [A\sin(\theta) + n_Q(t)]^2\}^{1/2} \quad (8.26)$$

As shown in the background section below, the probability density function for the envelope is given by

$$p(v) = \frac{v}{\sigma^2} \exp\left(\frac{A^2 + v^2}{-2\sigma^2}\right) I_0\left(\frac{Av}{\sigma^2}\right) \tag{8.27}$$

where $I_0(\cdot)$ is the modified zero-order Bessel function of the first kind, and σ^2 is the variance of the noise process $n(t)$. Since the envelope can be only nonnegative, Eq. (8.27) is identical to the pdf for a Rice random variable presented in Sec. 4.5. Equation (8.27) can be simplified by defining $z = v/\sigma$ and $a = A/\sigma$. Then the pdf can be rewritten in terms of the normalized variables z and a as:

$$p(z) = z \exp\left(\frac{a^2 + z^2}{-2}\right) I_0(az) \tag{8.28}$$

The power in the sine wave equals $A^2/2$, and the noise power equals σ^2; therefore, the signal-to-noise ratio is $A^2/(2\sigma^2) = a^2/2$. Notice that for $a = 0$, Eq. (8.28) reduces to the pdf for a Rayleigh random variable, which is the envelope distribution for bandpass gaussian noise in the absence of a signal.

Background. Defining

$$z_I(t) = A \cos(\theta) + n_I(t) \tag{8.29}$$

$$z_Q(t) = A \sin(\theta) + n_Q(t) \tag{8.30}$$

we can rewrite (8.25) as:

$$x(t) = z_I(t) \cos(\omega t) - z_Q(t) \sin(\omega t) \tag{8.31}$$

For any given value of θ, the components $z_I(t)$ and $z_Q(t)$ will be statistically independent gaussian random variables. The means and variances of $z_I(t)$ and $z_Q(t)$ are given by

$$E[z_I(t)] = A \cos \theta \tag{8.32}$$

$$E[z_Q(t)] = A \sin \theta \tag{8.33}$$

$$\operatorname{var}[z_I(t)] = E(\{z_I(t) - E[z_I(t)]\}^2)$$
$$= E[n_I^2(t)] = \sigma^2 \tag{8.34}$$

$$\operatorname{var}[z_Q(t)] = E(\{z_Q(t) - E[z_Q(t)]\}^2)$$
$$= E[n_Q^2(t)] = \sigma^2 \tag{8.35}$$

The joint pdf of $z_I(t)$ and $z_Q(t)$ conditioned on θ is then given by

$$p(z_I, z_Q \mid \theta) = \frac{1}{2\pi\sigma^2} \exp\left\{\frac{-1}{2\sigma^2}[(z_I - A\cos\theta)^2 + (z_Q - A\sin\theta)^2]\right\} \quad (8.36)$$

The envelope $v(t)$ and phase $\phi(t)$ are given by

$$v(t) = \sqrt{z_I^2(t) + z_Q^2(t)} \quad (8.37)$$

$$\phi(t) = \tan^{-1}\left(\frac{z_Q(t)}{z_I(t)}\right) \quad (8.38)$$

Following the procedure presented in Sec. 3.8, we can obtain the joint pdf for $v(t)$ and $\phi(t)$ from (8.36) as:

$$p(v, \phi \mid \theta) = \frac{v}{2\pi\sigma^2} \exp\left\{\frac{-1}{2\sigma^2}[v^2 + A^2 - 2Av\cos(\theta - \phi)]\right\} \quad (8.39)$$

Integration of (8.39) with respect to ϕ will yield the marginal pdf for v conditioned on θ:

$$\begin{aligned} p(v \mid \theta) &= \int_0^{2\pi} p(v, \phi \mid \theta)\, d\phi \\ &= \frac{v}{\sigma^2} \exp\left(\frac{A^2 + v^2}{-2\sigma^2}\right) I_0\left(\frac{Av}{\sigma^2}\right) \end{aligned} \quad (8.40)$$

Since the RHS of (8.40) is not a function of θ, it is actually an unconditional pdf as stated in (8.22).

8.9 Squared Envelope of Bandpass Gaussian Noise

In some applications it may be necessary to work with the squared envelope, which would be produced by a quadratic detector instead of the envelope produced by an envelope detector. The probability density function for the squared envelope is given by

$$p(z) = \frac{1}{2\sigma^2} \exp\left(\frac{-z}{2\sigma^2}\right) \quad z \geq 0 \quad (8.41)$$

This is the exponential distribution discussed in Sec. 4.3.

8.10 Phase of a Sine Wave Plus Bandpass Gaussian Noise [Refs. 1, 2, 4, 5, and 8]

Consider a signal:

$$x(t) = A\cos(\omega t + \theta) + n(t) \quad (8.42)$$

where $n(t)$ is a zero-mean bandpass gaussian noise process. It is possible to rewrite (8.42) in quadrature form as:

$$x(t) = [A\cos(\theta) + n_I(t)]\cos(\omega t) - [A\sin(\theta) + n_Q(t)]\sin(\omega t) \quad (8.43)$$

The instantaneous phase is then seen to be

$$\phi(t) = \tan^{-1}\left[\frac{A\sin\theta + n_Q(t)}{A\cos\theta + n_I(t)}\right] \quad (8.44)$$

For a particular value of θ, the phase ϕ has a probability density function given by

$$p(\phi\mid\theta) = \frac{\exp[-A^2/(2\sigma^2)]}{2\pi} + \frac{A\cos(\theta-\phi)}{2\sigma\sqrt{2\pi}}\exp\left[\frac{-A^2}{2\sigma^2}\sin^2(\theta-\phi)\right]$$

$$\cdot\left\{1 + \operatorname{erf}\left[\frac{A\cos(\theta-\phi)}{\sigma\sqrt{2}}\right]\right\} \quad (8.45)$$

Equation (8.45) is obtained by integrating Eq. (8.39) with respect to v. Denoting the signal-to-noise ratio by $R = A^2/(2\sigma^2)$, Eq. (8.45) can be

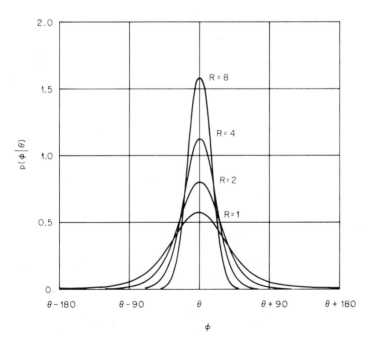

Figure 8.6 Probability density functions for the phase of a sine wave plus bandpass gaussian noise.

rewritten as:

$$p(\phi \mid \theta) = \frac{e^{-R}}{2\pi} + \frac{\sqrt{R}\cos(\theta-\phi)}{2\sqrt{\pi}} \exp[-R\sin^2(\theta-\phi)]$$
$$\cdot \{1 + \text{erf}\,[\sqrt{R}\cos(\theta-\phi)]\} \quad (8.46)$$

Plots of (8.46) for various values of R are shown in Fig. 8.6.

8.11 Postdetection Integration of Bandpass Gaussian Noise [Refs. 1 and 2]

It is often advantageous to sum up many independent outputs from a quadratic detector in order to form a "super sample" upon which to perform bit or symbol decisioning. Consider a system which sums k independent outputs from a quadratic detector:

$$\gamma = \sum_{i=1}^{k} z_i = \sum_{i=1}^{k} (I_i^2 + Q_i^2) \quad (8.47)$$

if each of the z_i are samples of the squared envelope for a bandpass gaussian noise process, the resulting output γ will have a probability density function given by

$$p(\gamma) = \frac{\gamma^{k-1}\exp(-\gamma/2\sigma^2)}{2^k \Gamma(k)\sigma^{2k}} \quad (8.48)$$

where σ^2 is the variance of the noise process. The distribution in (8.48) is the gamma distribution with $2k$ degrees of freedom.

8.12 REFERENCES

1. A. D. Whalen: *Detection of Signals in Noise*, Academic Press, New York, 1971.
2. H. Urkowitz: *Signal Theory and Random Processes*, Artech House, Dedham, Mass., 1983.
3. S. O. Rice: "Statistical Properties of a Sine Wave Plus Random Noise," *Bell System Technical Journal*, vol. 27, Jan. 1948, pp. 109–157.
4. S. O. Rice: "Mathematical Analysis of Random Noise (Part 1)," *Bell System Technical Journal*, vol. 23, July 1944, pp. 282–332.
5. S. O. Rice: "Mathematical Analysis of Random Noise (Part 2)," *Bell System Technical Journal*, vol. 24, Jan. 1945, pp. 46–156.
6. H. Taub and D. L. Schilling: *Principles of Communications Systems*, 2d ed., McGraw-Hill, New York, 1986.
7. S. Haykin: *Communication Systems*, 2d ed., Wiley, New York, 1983.
8. W. R. Bennett: "Methods of Solving Noise Problems," *Proc. IRE*, vol. 44, May 1956, pp. 609–638.
9. A. Papoulis: *Probability, Random Variables, and Stochastic Processes*, 2d ed., McGraw-Hill, New York, 1984.

10. G. A. Campbell and R. M. Foster: *Fourier Integrals for Practical Applications*, Van Nostrand, New York, 1954.
11. D. Knuth: *The Art of Computer Programming, Vol. 2: Semi-Numerical Algorithms*, 2d ed., Addison-Wesley, Reading, Mass., 1981.
12. G. R. Cooper and C. D. McGillem: *Modern Communication and Spread Spectrum*, McGraw-Hill, New York, 1986.
13. A. B. Carlson: *Communications Systems: An Introduction to Signals and Noise in Electrical Communication*, McGraw-Hill, New York, 1968.
14. J. G. Proakis: *Digital Communications*, McGraw-Hill, New York, 1983.

Chapter 9

Communication Channels

9.1 Radio Spectrum

By international agreement [Ref. 1], the radio frequency spectrum has been divided into bands with each band assigned a name as listed in Table 9.1. The types and severity of corruption that a signal will experience in traveling from the transmitter to the receiver will depend to a great extent upon the particular frequency band used. This chapter examines the types of signal corruption that may be experienced in the different bands and presents a few idealized mathematical models of this corruption that are commonly used for theoretically evaluating the performance of different signal waveforms and receiver designs. A second system of letter designations is often used for frequencies above 225 MHz (Table 9.2). According to Saveskie [Ref. 2], this system was established for security reasons by the military just prior to World War II. The system has survived to

TABLE 9.1 Radio Frequency Bands

Band		Frequency range
ELF:	Extremely low frequency	$30 < f \leq 300$ Hz
VF:	Voice frequency	$300 < f \leq 3000$ Hz
VLF:	Very low frequency	$3 < f \leq 30$ kHz
LF:	Low frequency	$30 < f \leq 300$ kHz
MF:	Medium frequency	$300 < f \leq 3000$ kHz
HF:	High frequency	$3 < f \leq 30$ MHz
VHF:	Very high frequency	$30 < f \leq 300$ MHz
UHF:	Ultrahigh frequency	$300 < f \leq 3000$ MHz
SHF:	Superhigh frequency	$3 < f \leq 30$ GHz
EHF:	Extremely high frequency	$30 < f \leq 300$ GHz

TABLE 9.2 Letter Designations for Frequency Bands above 225 MHz

Band	Frequencies, GHz	Band	Frequencies, GHz	Band	Frequencies, GHz
P	0.225 to 0.390	X_a	5.20 to 5.50	Q_a	36.0 to 38.0
L_p	0.390 to 0.465	X_q	5.50 to 5.75	Q_b	38.0 to 40.0
L_c	0.465 to 0.510	X_y	5.75 to 6.20	Q_c	40.0 to 42.0
L_l	0.510 to 0.725	X_d	6.20 to 6.25	Q_d	42.0 to 44.0
L_y	0.725 to 0.780	X_b	6.25 to 6.90	Q_e	44.0 to 46.0
L_t	0.780 to 0.900	X_r	6.90 to 7.00	V_a	46.0 to 48.0
L_s	0.900 to 0.950	X_c	7.00 to 8.50	V_b	48.0 to 50.0
L_x	0.950 to 1.150	X_l	8.50 to 9.00	V_c	50.0 to 52.0
L_k	1.150 to 1.350	X_s	9.00 to 9.60	V_d	52.0 to 54.0
L_f	1.350 to 1.450	X_x	9.60 to 10.00	V_e	54.0 to 56.0
L_z	1.450 to 1.550	X_f	10.00 to 10.25		
		X_k	10.25 to 10.90	W	56.0 to 100.0
S_e	1.55 to 1.65	K_p	10.90 to 12.25	C	3.90 to 6.20
S_f	1.65 to 1.85	K_s	12.25 to 13.25		
S_t	1.85 to 2.00	K_e	13.25 to 14.25	K_l	15.35 to 24.50
S_c	2.00 to 2.40	K_c	14.25 to 15.35		
S_q	2.40 to 2.60	K_u	15.35 to 17.25		
S_y	2.60 to 2.70	K_t	17.25 to 20.50		
S_g	2.70 to 2.90	K_q	20.50 to 24.50		
S_s	2.90 to 3.10	K_r	24.50 to 26.50		
S_a	3.10 to 3.40	K_m	26.50 to 28.50		
S_w	3.40 to 3.70	K_n	28.50 to 30.70		
S_h	3.70 to 3.90	K_l	30.70 to 33.00		
S_z	3.90 to 4.20	K_a	33.00 to 36.00		
S_d	4.20 to 5.20				

the present, but according to *Reference Data for Radio Engineers* [Ref. 5], the "designations have no official international standing."

9.2 Propagation

The study of electromagnetic propagation effects is a highly specialized subset of communication theory. Since this subject is treated at length in several dedicated texts [Refs. 2 through 4], the presentation here will be limited to an introduction to the various types of propagation and the effects that each can be expected to have upon a received signal. Detailed procedures for prediction of propagation performance are available in Saveskie, *Radio Propagation Handbook* [Ref. 2].

An *isotropic radiator* is a hypothetical point source of electromagnetic radiation that radiates equally well in all directions. A radio wave would propagate spherically from such a source, so at any given distance r, the power density would be equal to the total radiated power divided by the area of a sphere of radius r.

$$S = \frac{P}{4\pi r^2} \qquad (9.1)$$

The field strength provided by an isotropic radiator in free space provides a convenient benchmark for evaluating the relative performance of various types of propagation.

9.3 Ground Wave Propagation

As usually defined, the *ground wave* includes both the *surface wave* and the *space wave*. The space wave can be further separated into a direct component and a *terrestrially reflected* component.

Surface wave propagation [Refs. 2 and 3]

In *surface wave propagation*, the earth acts as a waveguide wall to guide an electromagnetic signal through the layer of atmosphere immediately above the surface of the earth. The thickness of the atmosphere layer in which the ground wave propagates varies from approximately 1 wavelength over dry land to several wavelengths over seawater. Transmissions in the ELF, VF, and lower VLF bands are propagated primarily via surface waves. Attenuation factors for various distances and frequencies in the VLF, LF, MF, and lower HF bands are listed in Tables 9.3 and 9.4. The attenuation factors are relative to the field strength at a distance of 1 km from an isotropic radiator. Examination of the tables indicates that the signal strength

TABLE 9.3 Ground Wave Attenuation over Dry Earth (Relative to Free-Space Attenuation at 1 km)

	Frequency							
km	10 kHz	60 kHz	150 kHz	500 kHz	1 MHz	2 MHz	5 MHz	10 MHz
1	0	0	0	0	−1.5	−11	−22.5	−29
2	−5.5	−5.5	−5.5	−5.5	−12	−22	−34	−41.5
5	−13.5	−13.5	−13.5	−16.5	−26.5	−38	−50	−57
10	−19	−19	−19	−27	−38.5	−50.5	−62	−70
20	−25.5	−25.5	−25.5	−38.5	−51	−62.5	−74	−82
50	−33	−33	−35	−54	−67.5	−79	−91.5	−100.5
100	−39	−40.5	−45	−67.5	−82	−93	−107	−118
200	−46	−48	−56.5	−83	−98.5	−112.5	−131	...
500	−55.5	−61	−77.5	−114	−135	...		
1000	−64	−74.5	−102	...				
2000	−77	−99	...					
5000	−111	...						

TABLE 9.4 Ground Wave Attenuation over Seawater (Relative to Free-Space Attenuation at 1 km)

km	\multicolumn{9}{c}{Frequency}
	10 kHz	60 kHz	150 kHz	500 kHz	1 MHz	2 MHz	5 MHz	10 MHz
1	0	0	0	0	0	0	0	0
2	−5.5	−5.5	−5.5	−5.5	−5.5	−5.5	−5.5	−5.5
5	−13.5	−13.5	−13.5	−13.5	−13.5	−13.5	−13.5	−13.5
10	−19	−19	−19	−19	−19	−19	−19	−19.5
20	−25.5	−25.5	−25.5	−25.5	−25.5	−25.5	−26	−27
50	−33	−33	−33	−33	−33	−33.5	−34.5	−38
100	−39	−39	−39	−39	−39.5	−42	−43	−48.5
200	−45.5	−45.5	−46	−47	−49.5	51.5	−55	−65
500	−54	−57	−59	−64	−67.5	−74	−84	−109
1000	−64	−71	−76	−86.5	−95	−107	−129	...
2000	−77	−92	−104	−128	...			
5000	−113	...						

decreases rapidly as distance and frequency increase. Seawater, being a better conductor than dry earth, makes a better waveguide as evidenced by the data in the tables. Additional data on surface wave propagation can be found in Ref. 4.

Space wave

The space wave consists of two components—the direct component that travels directly from the transmitting antenna to the receiving antenna, plus the terrestrially reflected component that reaches the receiving antenna after being reflected off the surface of the earth. This reflection causes a phase shift of approximately 180° in the reflected wave. At the receiving antenna, the phase of the direct component and the phase of the reflected component will differ by this amount plus an additional contribution due to the unequal path lengths traveled by the two components. At frequencies in the HF band and below, this second source of phase shift will be small, and the two components will be approximately 180° out of phase and thus tend to cancel each other. Space wave propagation is the predominant type of propagation in the UHF band.

Line-of-sight

In order for the direct component of the space wave to travel from the transmitting antenna to the receiving antenna, the two antennas must be within so-called *radio line-of-sight*. Due to bending of radio waves in the atmosphere, radio line-of-sight is not the same as optical line-of-sight. For this reason, some authors deprecate the use of *line-of-sight* (LOS) and propose alternatives such as *within radio horizon*. The nominal distance from an antenna to the radio horizon is given by

$$d = k\sqrt{h} \tag{9.2}$$

where h = height of the antenna

$$k = \begin{cases} 1.415 & \text{for } h \text{ in feet and } d \text{ in miles} \\ 4.124 & \text{for } h \text{ in meters and } d \text{ in kilometers} \end{cases}$$

Line-of-sight propagation between two antennas is theoretically possible only if

$$d_{TR} \leq d_{TH} + d_{RH} \tag{9.3}$$

where d_{TR} = distance between antennas
d_{TH} = distance from transmitting antenna to radio horizon
d_{RH} = distance from receiving antenna to radio horizon

Atmospheric conditions can cause actual LOS propagation ranges to be significantly longer or shorter than indicated by (9.3).

9.4 Additive Gaussian Noise Channel

The *additive gaussian noise* (AGN) channel is the simplest one of the several mathematical models commonly used in the analysis of communication systems. This model is based on the following assumptions:

1. The channel bandwidth is unlimited.
2. The channel attenuates the signal by a factor α that is assumed to be time-invariant and constant over all frequencies of interest.
3. The channel delays the signal by a constant amount t_0.
4. The channel adds zero-mean white gaussian noise to the signal.

If $x_{\text{LP}}(t)$ is a lowpass transmitted signal, the corresponding signal $y_{\text{LP}}(t)$ received via an AGN channel can be represented as:

$$y_{\text{LP}}(t) = \alpha x_{\text{LP}}(t - t_0) + n(t) \tag{9.4}$$

where α = attenuation factor
t_0 = time delay
$n(t)$ = a lowpass gaussian noise process

On the other hand, we could let $x(t)$ be a bandpass signal of the form:

$$x(t) = Re[\tilde{x}(t) \exp(j\omega_c t)] \tag{9.5}$$

where $\tilde{x}(t)$ is the complex envelope of $x(t)$ and ω_c is the carrier frequency in radians. The corresponding signal $y(t)$ received via an AGN channel can be represented as:

$$y(t) = Re[\tilde{y}(t) \exp(j\omega_c t)] \tag{9.6}$$

where $\tilde{y}(t) = \alpha \tilde{x}(t - t_0) \exp(-j\omega_c t_0) + \tilde{n}(t)$ (9.7)
α = attenuation factor
t_0 = time delay
ω_c = radian frequency of the carrier
$\omega_c t_0$ = carrier phase shift

The signal $\tilde{x}(t)$ is the equivalent lowpass signal (i.e., complex envelope) for $x(t)$, and $\tilde{n}(t)$ is a zero-mean gaussian random process that is the equivalent lowpass additive noise process corresponding to the bandpass noise process assumed to be operating in the channel.

It is customary to separate the issue of waveform performance from the issue of synchronization between the transmitter and receiver. Consequently, in the analysis of waveform performance, it will be convenient to eliminate from consideration the issue of synchronization by simply assuming that the time delay t_0 is known exactly at the receiver. Thus we can rewrite Eq. (9.7) as:

$$\tilde{y}(t) = \alpha \tilde{x}(t) \exp(-j\phi) + \tilde{n}(t) \tag{9.8}$$

where $\phi = \omega_c t_0$.

9.5 Band-Limited Channel

The assumptions inherent in the additive gaussian noise channel are inappropriate for many actual channels encountered in the real world. For modeling communications via wireline channels (such as telephone circuits), we must discard the assumption of unlimited channel bandwidth. The resulting channel model is commonly referred to as a *band-limited channel*.

If $x(t)$ is a bandpass signal having an equivalent lowpass signal $\tilde{x}(t)$, the corresponding lowpass equivalent signal $\tilde{y}(t)$ received via a band-limited channel can be represented as:

$$\tilde{y}(t) = \int_{-\infty}^{\infty} \tilde{x}(t) \tilde{h}(t - \tau) \, d\tau + n(t)$$

where $\tilde{h}(t)$ is the impulse response of the equivalent lowpass channel and $n(t)$ is a bandpass gaussian noise process. If the channel's amplitude response $|H(f)|$ is not constant for all frequencies within the bandwidth of the channel, the received signal will exhibit *amplitude distortion*. Furthermore, if the group delay characteristic (see Sec. 7.8) is not constant over the channel bandwidth, the received signal will exhibit *delay distortion*. The distortion present in signals transmitted via band-limited channels gives rise to the phenomenon of intersymbol interference (ISI).

9.6 References

1. *Radio Regulations of the International Telecommunication Union*, Article 2, Sec. 11, Geneva, 1959.
2. P. N. Saveskie: *Radio Propagation Handbook*, TAB Books, Blue Ridge Summit, Penn., 1980.
3. American Radio Relay League: *The ARRL Antenna Book*, Newington, Conn., 1974.
4. "Ground-wave Propagation Curves for Frequencies between 10 kHz and 30 MHz," Recommendation 368-4, CCIR, ITU, Geneva, 1982.
5. *Reference Data for Radio Engineers*, 6th ed., Howard W. Sams, Indianapolis, 1981.

Chapter

10

Detection Theory

Digital communication systems of various types can be represented by a model as shown in Fig. 10.1. The message source outputs a binary message consisting of a sequence of binary digits. Naturally, to be of any practical use, this sequence must somehow be related to the information to be conveyed. However, lacking a priori knowledge concerning this information, the rest of the system views the message sequence as being random in nature.

The transmitter outputs a signal $s_1(t)$ whenever the message bit is a 1, and a signal $s_0(t)$ whenever the message bit is a zero. As the transmitted signals travel to the receiver, they are corrupted by noise. In the model presented, this noise is represented as a random signal $n(t)$ which is added to the transmitted signal. In general, noise need not be additive in nature, but other characteristics of the corruption process tend to be much more difficult to work with. The additive assumptions have been successfully used to obtain extremely useful results in a great number of applications.

Ideally, the receiver should decode a 1 when $s_1(t)$ has been sent, and decode a zero when $s_0(t)$ has been sent. The heart of the problem lies in the fact that the receiver will "see" a signal which is either $s_1(t) + n(t)$ or $s_0(t) + n(t)$ and must decide whether in fact s_1 or s_0 was transmitted. If the noise corruption represented by $n(t)$ is sufficiently severe, the receiver may incorrectly decode 1 when s_0 was transmitted or decode zero when s_1 was transmitted. Typically, in the analysis of digital communication systems, both types of error are considered to be of equal importance. The total probability of error for both types is often specified as a *bit error rate* (BER), with values typically in the range 10^{-2} to 10^{-6}. A BER value of 10^{-3} means that *on the average* there will be a 1-bit error for every 1000 bits which are transmitted.

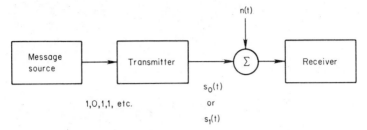

Figure 10.1 Model of a communication system.

Such BER specifications assume that the binary message sequence is completely random in nature. If some deterministic structure is incorporated into the message sequence, improvements in the BER performance are possible. However, the error performance of the channel without such added message structure is a necessary ingredient in order to compute overall performance of a system which does include some sort of error corrective coding.

10.1 Binary Decision Problem

General case

For the general communication system model shown in Fig. 10.1, there is a conceptual model of the corresponding binary decision problem shown in Fig. 10.2. Assume that the message source generates a sequence of outputs consisting of just H_0 and H_1. These hypotheses are mapped into points in the observation space by the transition mechanism. This mapping will be based on the two conditional probability densities $p(z \mid H_0)$ and $p(z \mid H_1)$. The former is the pdf of the observation z given that the active hypothesis is H_0, and the latter is the pdf of z given that the active hypothesis is H_1. The decision rule then operates on the coordinates of the points in the observation space to reach a decision concerning whether the source has generated H_0 or H_1. If the point lies in region R_0, then the rule will decide that H_0 was generated, and if the point lies in region R_1, the rule will decide H_1. Note that each region R_0 and R_1 need not be contiguous, but together they must cover the entire observation space. There are four possible combinations of events that can occur:

1. The source generates H_0 and the rule decides H_0.
2. The source generates H_0 and the rule decides H_1.
3. The source generates H_1 and the rule decides H_0.
4. The source generates H_1 and the rule decides H_1.

Detection Theory

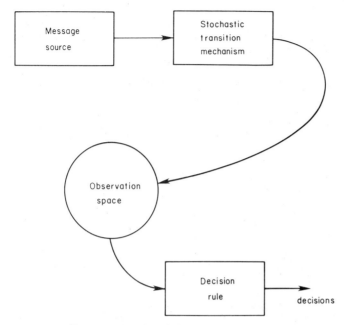

Figure 10.2 Conceptual model of the binary decision problem.

There are conditional probabilities associated with each of these four possibilities. Values for each can be obtained by integrating either $p(z \mid H_0)$ or $p(z \mid H_1)$ over the region R_0 or R_1 as appropriate and multiplying the result by the unconditional probability of the active hypothesis.

$$P(\text{decide } H_0 \mid H_0 \text{ true}) = P_0 \int_{R_0} p(z \mid H_0) \, dz \qquad (10.1)$$

$$P(\text{decide } H_1 \mid H_0 \text{ true}) = P_0 \int_{R_1} p(z \mid H_0) \, dz \qquad (10.2)$$

$$P(\text{decide } H_0 \mid H_1 \text{ true}) = P_1 \int_{R_0} p(z \mid H_1) \, dz \qquad (10.3)$$

$$P(\text{decide } H_1 \mid H_1 \text{ true}) = P_1 \int_{R_1} p(z \mid H_1) \, dz \qquad (10.4)$$

where P_0 = absolute probability that hypothesis H_0 is correct
P_1 = absolute probability that hypothesis H_1 is correct

Binary decisions using a single decision variable

In typical binary decision problems, the receiver must compare the value of a received parameter to a threshold and attempt to determine which one of two possible signals has been transmitted. Often there is actually only one possible signal, and the receiver must decide whether or not it has been sent. Noise can corrupt the value of the received parameter so that it appears to the receiver as though signal s_1 has been sent when in fact s_2 was sent or vice versa. Consider the following case:

$$H_0: s_0(t) = 0 \quad z(t) = n(t)$$

$$H_1: s_1(t) = s \quad z(t) = s + n(t)$$

where $n(t)$ is a zero-mean gaussian process and s is a constant. The conditional pdf's for z given H_0 and H_1 are shown in Fig. 10.3 along with an arbitrary threshold value of $z = \lambda$. If the value of z observed by the receiver is less than λ, the receiver will decide that hypothesis H_0 is true. Conversely, if the observed value of z is greater than λ, the receiver will decide that hypothesis H_1 is true. Various strategies for optimal selection of λ are discussed in Sec. 10.2.

If hypothesis H_0 is true but the observed value of z exceeds λ, the receiver will incorrectly decide that hypothesis H_1 is true. This is called a *false alarm* or *false detection*. The probability of a false alarm is equal to the area under that part of $p(z \mid H_0)$ which lies to the right of $z = \lambda$ (that is, the shaded area in Fig. 10.4a).

$$P_F = \int_\lambda^\infty p(z \mid H_0)\, dz = P(d_1 \mid H_0) \tag{10.5}$$

The probability of *miss* or *false dismissal* is equal to the area under

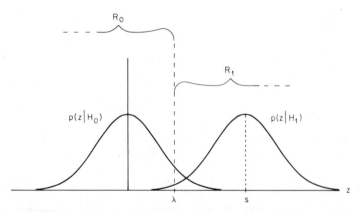

Figure 10.3 Conditional probability density functions for a binary decision problem.

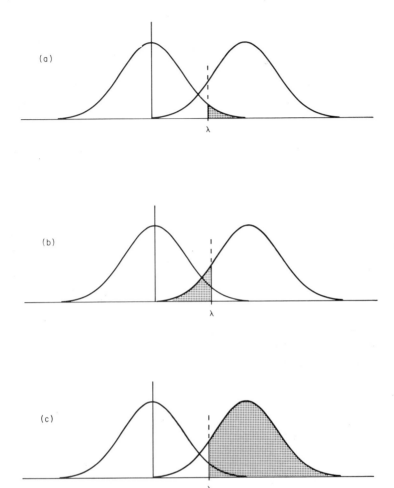

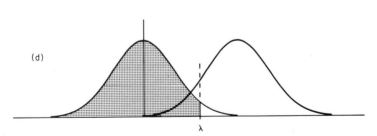

Figure 10.4 Probability density functions for a binary decision problem. Shaded areas represent (a) probability of false alarm, (b) probability of false dismissal, (c) probability of detection, and (d) probability of correct dismissal.

that part of $p(z\,|\,H_1)$ which lies to the left of $z = \lambda$.

$$P_M = \int_{-\infty}^{\lambda} p(z\,|\,H_1)\,dz = P(d_0\,|\,H_1) \qquad (10.6)$$

The probability of detection is equal to the area under that part of $p(z\,|\,H_1)$ which lies to the right of $z = \lambda$.

$$P_D = \int_{\lambda}^{\infty} p(z\,|\,H_1)\,dz = P(d_1\,|\,H_1) \qquad (10.7)$$

The probability of correct dismissal is equal to the area under that part of $p(z\,|\,H_0)$ which lies to the left of $z = \lambda$.

$$P_A = \int_{-\infty}^{\lambda} p(z\,|\,H_0)\,dz = P(d_0\,|\,H_0) \qquad (10.8)$$

10.2 Optimal Decision Criteria

A number of different criteria have been developed for optimal selection of the threshold in binary decision problems. None of these criteria is inherently "better" than the others; rather, one criterion may be more appropriate than all others given the specific circumstances surrounding any particular application.

Maximum likelihood decision criterion

The maximum likelihood (ML) criterion is the simplest useful decision criterion, and it is inadequate for many applications. Simply put, the ML criterion chooses the hypothesis which is the most likely to have caused the observed value of z. The likelihood ratio test for the maximum likelihood criterion is given by

$$\Lambda(z) = \frac{p(z\,|\,H_1)}{p(z\,|\,H_0)} \underset{H_0}{\overset{H_1}{\gtrless}} 1 \qquad (10.9)$$

An equivalent way of stating the same decision rule is given by

Choose H_0 if $p(z\,|\,H_0) > p(z\,|\,H_1)$

Choose H_1 if $p(z\,|\,H_1) > p(z\,|\,H_0)$

The ML criterion is equivalent to the ideal observer criterion with $P(H_0) = P(H_1)$.

Bayes decision criterion

In the Bayes decision criterion, the designer assigns a cost to each possible correct decision and incorrect decision and then sets the threshold of the likelihood ratio test such that the average cost is minimized. The likelihood ratio test for the Bayes criterion is given by

$$\frac{p(z \mid H_1)}{p(z \mid H_0)} \underset{H_0}{\overset{H_1}{\gtrless}} \frac{(C_{10} - C_{00})P(H_0)}{(C_{01} - C_{11})P(H_1)} \qquad (10.10)$$

where $C_{nk} \triangleq$ cost of making decision d_n given that hypothesis H_k is true
$P(H_n) \triangleq$ the a priori probability that hypothesis H_n is true
$p(z \mid H_n) \triangleq$ the conditional pdf of the observation z given that hypothesis H_n is true

Ideal observer decision criterion

The *ideal observer* criterion is also called the *probability of error* criterion, or the *minimum error* criterion, and is equivalent to the *maximum a posteriori* (MAP) criterion. The likelihood ratio test for the ideal observer criterion is given by

$$\Lambda(z) = \frac{p(z \mid H_1)}{p(z \mid H_0)} \underset{H_0}{\overset{H_1}{\gtrless}} \frac{P(H_0)}{P(H_1)} \qquad (10.11)$$

Note that this corresponds to the Bayes criterion with $C_{11} = C_{00} = 0$ and $C_{10} = C_{01}$.

Neyman-Pearson decision criterion

The Neyman-Pearson criterion allows the designer to specify a maximum value α for the probability of false alarm P_F. The Neyman-Pearson criterion then sets the threshold to maximize the probability of detection P_D for $P_F = \alpha$. The likelihood ratio test for the Neyman-Pearson criterion is given by

$$\Lambda(z) = \frac{p(z \mid H_1)}{p(z \mid H_0)} \underset{H_0}{\overset{H_1}{\gtrless}} \lambda \qquad (10.12)$$

The threshold λ is found using

$$P_F = \alpha = \int_{R_1} p(z \mid H_0) \, dz \qquad (10.13)$$

The integration over R_1 will involve λ in one of the limits of integration. Equation (10.13) can thus be solved for λ.

Example Given:

$$p(z \mid H_0) = \frac{1}{\sqrt{2\pi}} e^{-z^2/2}$$

$$p(z \mid H_1) = \frac{1}{\sqrt{2\pi}} e^{-(z-2)^2/2}$$

$$P_F = \alpha = 0.4$$

We can then form the Neyman-Pearson likelihood ratio test:

$$\Lambda(z) = \frac{(1/\sqrt{2\pi})e^{-(z-2)^2/2}}{(1/\sqrt{2\pi})e^{-z^2/2}} = e^{2z-2} \underset{H_0}{\overset{H_1}{\gtrless}} \lambda$$

Taking the natural logarithm of both sides yields

$$2z - 2 \underset{H_0}{\overset{H_1}{\gtrless}} \ln \lambda$$

$$z \underset{H_0}{\overset{H_1}{\gtrless}} \frac{\ln \lambda + 2}{2}$$

The threshold λ can now be obtained using

$$P_F = \alpha = 0.4 = \int_{1+(\ln \lambda/2)}^{\infty} \frac{1}{\sqrt{2\pi}} e^{-z^2/2} \, dz$$

$$Q(1 + 0.5 \ln \lambda) = 0.4$$

Using the table of Q function values, we obtain

$$1 + 0.5 \ln \lambda = 0.25$$

$$\lambda = 0.223$$

10.3 Optimum Coherent Detection of Binary Signals in the AWGN Channel

A matched-filter receiver structure for optimum coherent detection of binary signals is shown in Fig. 10.5. This receiver is optimum in the sense that it uses the *maximum posterior probability* criterion, that is,

$$P[s_1 \mid r(t)] \underset{d=b_2}{\overset{d=b_1}{\gtrless}} P[s_2 \mid r(t)]$$

where $P[s_1 \mid r(t)]$ is the posterior probability—i.e., the probability that s_1 was transmitted given that $r(t)$ was received. It is assumed that the two signals are equally likely and have equal energy. The inputs and outputs of the matched filters are related via

$$z_n(t) = \int_{-\infty}^{\infty} r(\tau) s_n(T - t + \tau) \, d\tau \qquad n = 1, 2 \qquad (10.14)$$

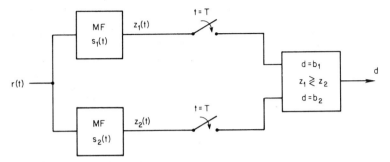

Figure 10.5 Matched-filter receiver for detection of binary signals.

At the sampling instant $t = T$, Eq. (10.14) reduces to

$$z_n(t) = \int_{-\infty}^{\infty} r(\tau)s_n(\tau)\,d\tau \qquad n = 1, 2 \qquad (10.15)$$

Since $s_n(t)$ is defined to be zero outside of the interval $0 \le t \le T$, the limits of integration can be changed to yield

$$z_n(t) = \int_0^T r(\tau)s_n(\tau)\,d\tau \qquad n = 1, 2 \qquad (10.16)$$

Equation (10.16) is in the form of a correlation and suggests an alternative receiver structure as shown in Fig. 10.6. When dealing with bandpass signals, it will often be more convenient to analyze receiver structures that are formulated in terms of the equivalent lowpass signals. Assume that the transmitted signals are of the form given by Eq. (9.5), and the corresponding received signals are of the form given by Eq. (9.6). The corresponding matched-filter receiver and correlation receiver are shown in Figs. 10.7 and 10.8. For these

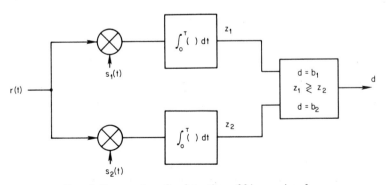

Figure 10.6 Correlation receiver for detection of binary signals.

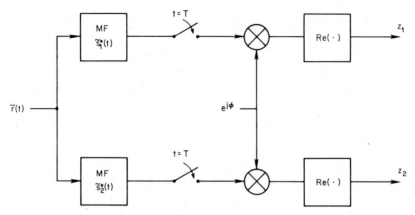

Figure 10.7 Matched-filter receiver for detection of bandpass binary signals in terms of lowpass equivalent signals.

receivers, the decision variables are given by

$$z_1 = Re\left[\exp(j\phi) \int_0^T \tilde{r}(t)\tilde{s}_1^*(t)\, dt \right] \qquad (10.17)$$

$$z_2 = Re\left[\exp(j\phi) \int_0^T \tilde{r}(t)\tilde{s}_2^*(t)\, dt \right] \qquad (10.18)$$

When $s_1(t)$ is transmitted, the complex envelope of the received signal is given by

$$\tilde{r}(t) = \alpha \exp(-j\phi)s_1(t) + \tilde{n}(t) \qquad (10.19)$$

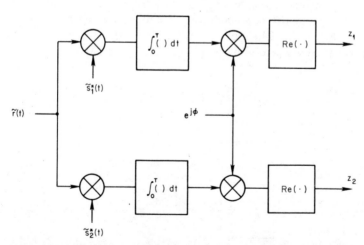

Figure 10.8 Correlation receiver for detection of bandpass binary signals in terms of lowpass equivalent signals.

and the decision variables become

$$z_1 = 2\alpha E + N_1 \tag{10.20}$$

$$z_2 = 2\alpha \rho E + N_2 \tag{10.21}$$

where $N_m = Re\left[\exp(j\phi) \int_0^T \tilde{n}(t)\tilde{s}_m^*(t)\, dt\right]$ (10.22)

$$E = \int_0^T s_1^2(t)\, dt = \int_0^T s_2^2(t)\, dt$$

$$= \frac{1}{2}\int_0^T |\tilde{s}_1(t)|^2\, dt = \frac{1}{2}\int_0^T |\tilde{s}_2(t)|^2\, dt \tag{10.23}$$

The noise terms N_1 and N_2 are jointly gaussian, zero-mean random variables. For a given signal $s(t)$, the terms $2\alpha E$ and $2\alpha \rho E$ are nonrandom. Thus z_1 is a gaussian random variable with a mean of $2\alpha E$, and z_2 is gaussian with a mean of $2\alpha \rho E$. Comparing z_1 to z_2 is equivalent to comparing their difference to zero. The difference $(z_2 - z_1)$ is a gaussian random variable with a mean of $2\alpha E(1-\rho)$ and variance of $4EN_0(1-\rho)$.

Given that $s_1(t)$ is transmitted, the probability of error is given by

$$P(z_2 > z_1 \mid s_1) = P[(z_2 - z_1) > 0 \mid s_1] = \frac{1}{2}\,\mathrm{erfc}\!\left(\sqrt{(1-\rho)\frac{\alpha^2 E}{2N_0}}\right) \tag{10.24}$$

where ρ is the correlation coefficient:

$$\rho = \frac{1}{\sqrt{E_1 E_2}} \int_0^T s_1(t) s_2(t)\, dt$$

$$= Re\left[\frac{1}{2\sqrt{E_1 E_2}} \int_0^T \tilde{s}_1(t)\tilde{s}_2^*(t)\, dt\right]$$

Given that $s_2(t)$ is transmitted, the probability $P(z_1 > z_2)$ is identical to (10.24). Therefore, since s_1 and s_2 were assumed to be equally likely, we have for the unconditional probability of error:

$$P_e = P(z_2 > z_1) = P(z_1 > z_2)$$

As indicated in (10.24), the probability of error is a function of the correlation between s_1 and s_2. The special cases of antipodal signals and orthogonal signals are important enough to warrant special mention. The probability of error as given by (10.24) is plotted in Fig. 10.9 as a function of E_b/N_0 for several values of ρ.

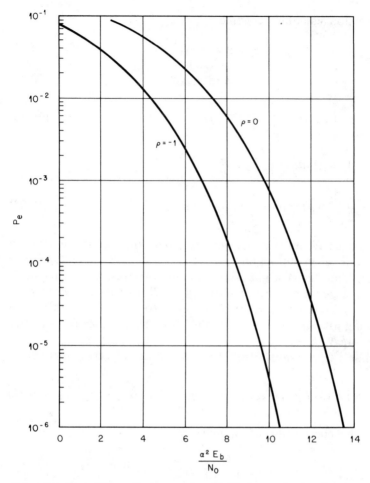

Figure 10.9 Probability of error versus E_b/N_0 for optimal coherent detection of binary signals in AWGN.

Antipodal signals

When a binary signal set exhibits the property

$$s_1(t) = -s_2(t) \qquad (10.25)$$

$$\tilde{s}_1(t) = -\tilde{s}_2(t) \qquad (10.26)$$

the signals are said to be *antipodal*. [*Note*: Equation (10.25) implies (10.26), and vice versa.] For antipodal signals, the correlation coefficient ρ is equal to -1, thus maximizing the argument of the complementary error function (10.24) to yield

$$P_e = \frac{1}{2}\text{erfc}\left(\sqrt{\frac{\alpha^2 E}{N_0}}\right) \tag{10.27}$$

The term under the radical in (10.27) is just the received SNR per bit commonly denoted as E_b/N_0.

Orthogonal signals

A binary signal set is said to be *orthogonal* when the correlation coefficient is equal to zero. For orthogonal signals, the probability of error given by (10.24) reduces to

$$P_e = \frac{1}{2}\text{erfc}\left(\sqrt{\frac{\alpha^2 E}{2N_0}}\right) \tag{10.28}$$

Orthogonal signals require twice as much (that is, 3 dB more) energy per bit as antipodal signals require to achieve equal error performance.

10.4 Optimum Coherent Detection of *M*-ary Signals in the AWGN Channel

The receiver structure for optimum coherent demodulation of binary signals presented in the previous section can be extended to the *M*-ary signal case. It is assumed that the set of transmitted signals is given by

$$s_m(t) = Re[\tilde{s}_m(t)\exp(j\omega_c t)] \quad m = 1, 2, \ldots, M \tag{10.29}$$

It is further assumed that each of the transmitted signals has equal energy.

Based upon Eq. (9.5), the receiver input $r(t)$ is represented as:

$$r(t) = Re[\tilde{r}(t)\exp(j\omega_c t)] \tag{10.30}$$

where the complex envelopes or equivalent lowpass signals are given by

$$\tilde{r}(t) = \alpha\tilde{s}_m(t)\exp(-j\phi) + \tilde{n}(t) \tag{10.31}$$

The decision variables are given by

$$z_m = Re\left[\exp(j\phi)\int_0^T \tilde{r}(t)\tilde{s}_m^*(t)\,dt\right] \tag{10.32}$$

When $s_1(t)$ is transmitted, the complex envelope of the received signal is given by

$$\tilde{r}(t) = \alpha \exp(-j\phi)s_1(t) + \tilde{n}(t)$$

and the decision variable z_1 is exactly as Eq. (10.20).

$$z_1 = 2\alpha E + N_1 \tag{10.33}$$

The decision variables z_m for $m = 2, 3, \ldots, M$ depend upon what is assumed about the cross-correlations between s_1 and the remaining s_m.

Orthogonal signals

Assume that the M signals are all mutually orthogonal. Under this assumption, given that s_1 is transmitted, the decision variable z_1 is gaussian-distributed with a mean of $2\alpha E$, a variance of $2EN_0$, and a pdf given by

$$p(z_1 \mid s_1) = \frac{1}{2\sqrt{E\pi N_0}} \exp\left(\frac{-(z_1 - 2\alpha E)^2}{4EN_0}\right)$$

The remaining decision variables are zero-mean gaussian RVs with a variance of $2EN_0$ and pdf's given by

$$p(z_m \mid s_1) = \frac{1}{2\sqrt{E\pi N_0}} \exp\left(\frac{-z_m^2}{4EN_0}\right) \quad m = 2, 3, \ldots, M$$

Using this information, it can be shown [Ref. 1] that the probability of making a symbol error is given by

$$P_s = \frac{1}{\sqrt{2\pi}} \int_{-\infty}^{\infty} \left(1 - \left[1 - \frac{1}{2}\mathrm{erfc}\left(\frac{y}{\sqrt{2}}\right)\right]^{M-1}\right)$$

$$\cdot \exp\left[\frac{-(y - \sqrt{2kE_b/N_0})^2}{2}\right] dy \tag{10.34}$$

where $E_b = \alpha^2 E/k$ is the received energy per bit. An alternative form for P_s is given by

$$P_s = 1 - \frac{1}{\sqrt{2\pi}} \int_{-\infty}^{\infty} \exp\left(\frac{-y^2}{2}\right)\left[1 - \frac{1}{2}\mathrm{erfc}\left(\frac{y}{\sqrt{2}} + \sqrt{\frac{kE_b}{N_0}}\right)\right]^{M-1} dy \tag{10.35}$$

One symbol error can result in one or more bit errors. The symbol error probability given in (10.34) or (10.35) can be converted into an

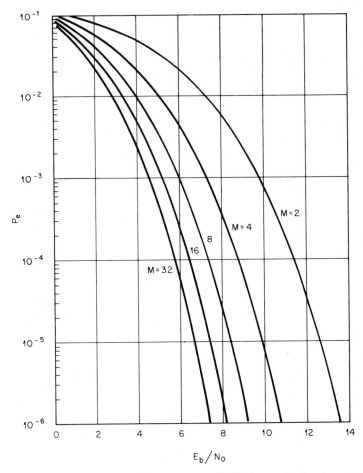

Figure 10.10 Probability of error versus E_b/N_0 for optimal coherent detection of orthogonal M-ary signals in AWGN.

average bit error probability P_e, using

$$P_e = \frac{2^{k-1}}{2^k - 1} P_s \qquad (10.36)$$

where $K = \log_2 M$

Plots of P_e versus E_b/N_0 for several values of M are shown in Fig. 10.10.

Biorthogonal signals

An M-ary biorthogonal signal set consists of $M/2$ pairs of antipodal signals. Furthermore, each signal is orthogonal to all signals in the set other than itself and its antipodal mate. An example of a biorthogonal signal is QPSK with phase shifts of 0°, 90°, 180°, and 270°. The signals shifted by 0° and 90° form an antipodal pair as do the signals shifted by 180° and 270°. Also each signal is orthogonal to the two signals in the other pair. An M-ary biorthogonal signal set requires a bandwidth equal to the bandwidth needed by a set of $M/2$ orthogonal waveforms or one-half the bandwidth required by a set of M orthogonal waveforms. An optimum receiver structure for detection of biorthogonal signals is shown in Fig. 10.11. The detection process can be viewed as an optimum N-ary detection ($N = M/2$) to decide upon one of the $M/2$ antipodal pairs, followed by an optimum binary detection on that pair to decide upon one of the two antipodal signals comprising the pair. The probability of symbol error is given by Refs. 2 and 3 as:

$$P_s = 1 - \frac{1}{\sqrt{2\pi}} \int_{-\sqrt{2\gamma}}^{\infty} \exp\left(\frac{-y^2}{2}\right) \left[\frac{1}{\sqrt{2\pi}} \int_{-y-\sqrt{2\gamma}}^{y+\sqrt{2\gamma}} \exp\left(\frac{-x^2}{2}\right) dx\right]^{(M/2)-1} dy$$

(10.37)

where $\gamma = kE_b/N_0$. Plots of P_s versus E_b/N_0 for several values of M are shown in Fig. 10.12.

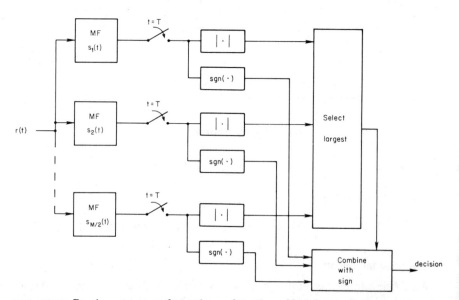

Figure 10.11 Receiver structure for optimum detection of biorthogonal signals.

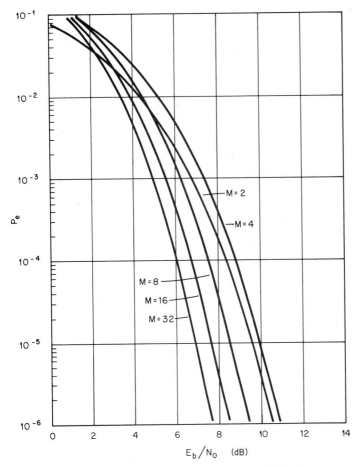

Figure 10.12 Probability of symbol error versus E_b/N_0 for optimal detection of biorthogonal signals in AWGN.

Equicorrelated signals

An *equicorrelated* or *transorthogonal* signal set consists of M different signal waveforms where the cross-correlations between all possible pairs of waveforms are equal. It can be shown that the probability of symbol error for optimum detection of such a signal set is given by

$$P_s = 1 - \frac{1}{\sqrt{2\pi}} \int_{-\infty}^{\infty} \exp\left(\frac{-y^2}{2}\right) \left[\frac{1}{\sqrt{2\pi}} \int_{-\infty}^{y+\sqrt{2\gamma(1-\rho)}} \exp\left(\frac{-x^2}{2}\right) dx\right]^{M-1} dy \tag{10.38}$$

where ρ is the correlation between pairs of waveforms in the signal set. It can be shown [Refs. 1 through 3] that

$$\rho \geq \frac{-1}{M-1} \qquad (10.39)$$

When the equality in (10.39) is satisfied, the signal set is called a *simplex* signal set. (Some texts such as Ref. 4 use the term *simplex* even when ρ is not at the minimum.) It is believed that the simplex set is the optimum set of M-ary waveforms for the AWGN channel [Ref. 5]. A simplex set of waveforms can achieve any given level of error performance at an SNR that is only $(m-1)/M$ times the SNR required by an orthogonal set of waveforms for the same level of

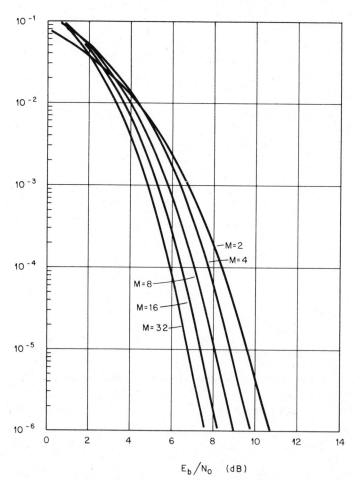

Figure 10.13 Probability of symbol error versus E_b/N_0 for optimal detection of simplex signals in AWGN.

performance. The savings are most dramatic for small M and become insignificant for large M. Plots of P_s versus E_b/N_0 for simplex signal sets are shown in Fig. 10.13.

10.5 Optimum Noncoherent Detection in an AWGN Channel

Binary signals

Let the two signal waveforms of a binary signal set be represented by

$$s_1(t) = Re[\tilde{s}_1(t) \exp(j\omega_c t)]$$

$$s_2(t) = Re[\tilde{s}_2(t) \exp(j\omega_c t)]$$

Assume that the two signals are equally likely and have equal energy E. Their complex-valued correlation coefficient is given by

$$\rho = \frac{1}{2E} \int_0^T \tilde{s}_1(t)\tilde{s}_2(t)\, dt \qquad (10.40)$$

A receiver structure for optimum noncoherent detection of the signals is shown in Fig. 10.14. It can be shown that the probability of error is given by

$$P_e = Q(a, b) - \frac{1}{2} I_0(ab) \exp\left(-\frac{a^2 + b^2}{2}\right) \qquad (10.41)$$

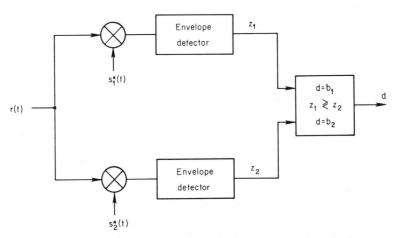

Figure 10.14 Receiver structure for optimum noncoherent detection of binary signals in AWGN.

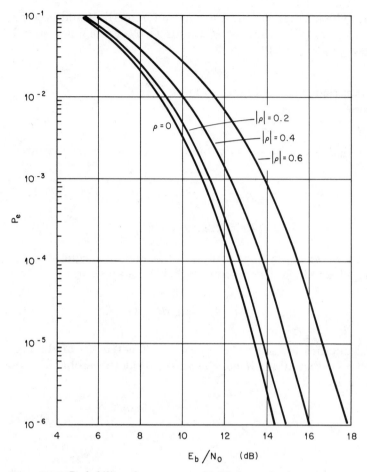

Figure 10.15 Probability of error versus E_b/N_0 for optimal noncoherent detection of binary signals in AWGN.

where
$$a = \left[\frac{E_b}{2N_0}(1 - \sqrt{1 - |\rho|^2})\right]^{1/2}$$

$$b = \left[\frac{E_b}{2N_0}(1 + \sqrt{1 - |\rho|^2})\right]^{1/2}$$

$Q(\cdot, \cdot)$ is the Marcum Q function (Sec. 2.8)
$I_0(\cdot)$ is the modified Bessel function of order zero (Sec. 2.7)

The probability of error P_e is plotted in Fig. 10.15 for several values of $|\rho|$.

M-ary orthogonal signals

The binary receiver structure shown in Fig. 10.14 can be extended to the M-ary case. For an M-ary set of orthogonal waveforms, the probability of symbol error is given by

$$P_s = \sum_{n=1}^{M-1} \frac{(-1)^{n+1}(M-1)!}{(n+1)!(M-1-n)!} \exp\left(\frac{-nkE_b}{(n+1)N_0}\right) \qquad (10.42)$$

10.6 References

1. J. G. Proakis: *Digital Communications*, McGraw-Hill, New York, 1983.
2. S. Stein and J. J. Jones: *Modern Communications Principles*, McGraw-Hill, New York, 1967.
3. A. H. Nuttall: "Error Probabilities for Equicorrelated M-ary Signals under Phase Coherent and Phase Incoherent Reception," *IRE Trans. Information Theory*, IT-8, July 1962, pp. 305–314.
4. G. R. Cooper and C. D. McGillem: *Modern Communications and Spread Spectrum*, McGraw-Hill, New York, 1986.
5. A. J. Viterbi: *Principles of Coherent Communication*, McGraw-Hill, New York, 1966.

Chapter

11

Signal Processing and Simulation Issues

11.1 Digital Processing in Communication Systems

Digital signal processing (DSP) is based on the fact that an analog signal can be digitized and input to either a general-purpose digital processor or a special-purpose digital processor. All sorts of mathematical operations may then be performed upon the sequence of digital values inside the processor. Some of these operations are simply digital versions of classical analog techniques, while others have no counterpart in analog circuit devices or processing methods. Digital signal processing techniques play a central role in many types of modern, state-of-the-art communications equipment.

Digitization

Digitization is the process of converting an analog signal such as a time-varying voltage or current into a sequence of digital values. Digitization actually involves two distinct parts—*sampling* and *quantization*—which are usually analyzed separately for the sake of convenience and simplicity. Three basic types of sampling, shown in Fig. 11.1, are *ideal sampling, instantaneous sampling*, and *natural sampling*. From the illustration we can see that the sampling process converts a signal that is defined over a continuous time interval into a signal which has nonzero amplitude values only at discrete instants of time (as in ideal sampling) or over a number of discretely separate but internally continuous subintervals of time (as in instantaneous and natural sampling). The signal which results from a sampling

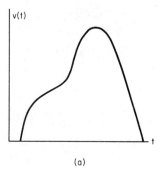
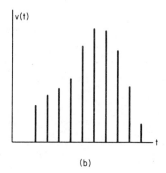
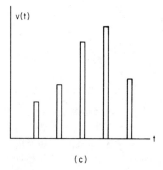
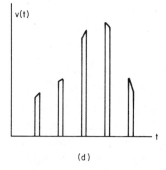

Figure 11.1 An analog signal (*a*) and three different types of sampling: (*b*) ideal, (*c*) instantaneous, and (*d*) natural.

process is called a *sampled-data* signal. The signals resulting from ideal sampling are also referred to as *discrete-time* signals.

Each of the three basic sampling types occurs at different places within a DSP system. The output from a sample-and-hold amplifier or a digital-to-analog converter (DAC) is an instantaneously sampled signal. The sequence of discrete values produced by an analog-to-digital converter (ADC) can be viewed as either an instantaneously sampled signal or an ideally sampled signal, but the latter is almost universally preferred. Natural sampling is encountered in the analysis of the analog multiplexing that is often performed prior to A/D conversion in multiple-signal systems. In all three of the sampling approaches presented, the sample values are free to assume any appropriate value from the continuum of possible analog signal values.

Quantization is the part of digitization which is concerned with converting the amplitudes of an analog signal into values which can be represented by binary numbers having some finite number of bits.

Signal Processing and Simulation Issues   165

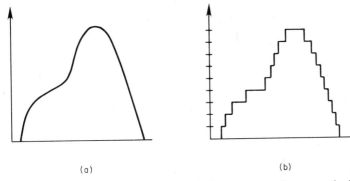

(a)   (b)

Figure 11.2 An analog signal (a) and the corresponding quantized signal (b).

A quantized or *discrete-valued* signal is shown in Fig. 11.2. The sampling and quantization processes will introduce some significant changes in the spectrum of a digitized signal. The details of the changes will depend upon both the precision of the quantization operation and the particular sampling model that most aptly fits the actual situation.

Ideal sampling

In *ideal sampling*, the sampled-data signal, as shown in Fig. 11.3, comprises a sequence of uniformly spaced impulses, with the weight of each impulse equal to the amplitude of the analog signal at the corresponding instant in time. Although not mathematically rigorous, it is convenient to think of the sampled-data signal as the result of multiplying the analog signal $x(t)$ by a periodic train of unit impulses.

$$x_s(\,\cdot\,) = x(t) \sum_{n=-\infty}^{\infty} \delta(t - nT)$$

Based upon property 11 from Table 6.2, this means that the spectrum of the sampled-data signal could be obtained by convolving the

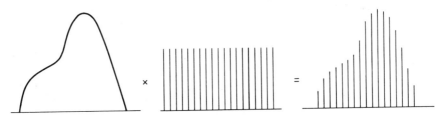

Figure 11.3 Ideal sampling.

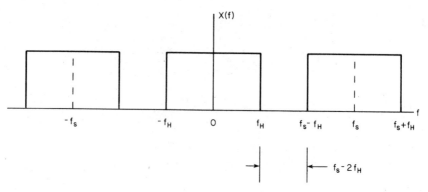

Figure 11.4 Spectrum of an ideally sampled signal.

spectrum of the analog signal with the spectrum of the impulse train.

$$\mathscr{F}\left[x(t)\sum_{n=-\infty}^{\infty}\delta(t-nT)\right]=X(f)\otimes\left[f_s\sum_{m=-\infty}^{\infty}\delta(f-mf_s)\right]$$

As illustrated in Fig. 11.4, this convolution produces copies, or *images*, of the original spectrum that are periodically repeated along the frequency axis. Each of the images is an exact copy of the original spectrum. The center-to-center spacing of the images is equal to the sampling rate f_s, and the edge-to-edge spacing is equal to $f_s - 2f_H$. As long as f_s is greater than 2 times f_H, the original signal can be recovered by a lowpass filtering operation which removes the extra images introduced by the sampling.

Sampling rate selection

If f_s is less than $2f_H$, the images will overlap or *alias* as shown in Fig. 11.5, and recovery of the original signal will not be possible. The minimum alias-free sampling rate of $2f_H$ is called the *Nyquist rate*. A signal sampled exactly at its Nyquist rate is said to be *critically sampled*.

Uniform Sampling Theorem

If the spectrum $X(f)$ of a function $x(t)$ vanishes beyond an upper frequency of f_H Hz or ω_H rad/sec (radians per second), then $x(t)$ can be completely determined by its values at uniform intervals of less than $1/(2f_H)$ or π/ω. If sampled within these constraints, the original function $x(t)$ can be reconstructed from the samples by

$$x(t)=\sum_{n=-\infty}^{\infty}x(nT)\frac{\sin[2f_s(t-nT)]}{2f_s(t-nT)}$$

where T is the sampling interval.

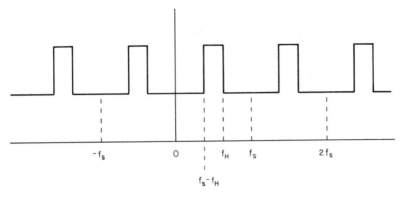

Figure 11.5 Aliasing due to overlap of spectral images.

Since practical signals cannot be strictly band-limited, sampling of a real-world signal must be performed at a rate greater than $2f_H$ where the signal is known to have negligible (i.e., typically less than 1 percent) spectral energy above the frequency of f_H. When designing a signal processing system, we will rarely, if ever, have reliable information concerning the exact spectral occupancy of the noisy real-world signals that our system will eventually face. Consequently, in most practical design situations, a value is selected for f_H based upon the requirements of the particular application, and then the signal is lowpass filtered prior to sampling. Filters used for this purpose are called *antialiasing*, or *guard*, filters. The sample-rate selection and guard-filter design are coordinated so that the filter provides attenuation of 40 dB or more for all frequencies above $f_s/2$. The spectrum of an ideally sampled practical signal is shown in Fig. 11.6. Although some aliasing does occur, the aliased components are suppressed at least 40 dB below the desired components. Antialias filtering must be performed prior to sampling. In general, there is no way to eliminate aliasing once a signal has been improperly sampled. The particular type (Butterworth, Chebyshev, Bessel, Cauer, etc.), and order of the filter should be chosen to provide the necessary stopband attenuation while preserving the passband characteristics most important to the intended application.

Instantaneous sampling

In instantaneous sampling, each sample has a nonzero width and a flat top. As shown in Fig. 11.7, the sampled-data signal resulting from instantaneous sampling can be viewed as the result of convolving a sample pulse $p(t)$ with an ideally sampled version of the analog

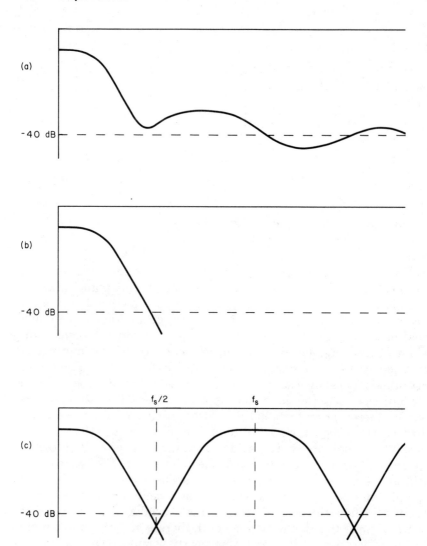

Figure 11.6 Spectrum of an ideally sampled practical signal. (*a*) Spectrum of raw analog signal. (*b*) Spectrum after lowpass filtering. (*c*) Spectrum after sampling.

signal. The resulting sampled-data signal can thus be expressed as:

$$x_s(\,\cdot\,) = p(t) \otimes \left[x(t) \sum_{n=-\infty}^{\infty} \delta(t-nT) \right]$$

where $p(t)$ is a single rectangular sampling pulse and $x(t)$ is the original analog signal. Based upon property 10 from Table 6.2, this means that the spectrum of the instantaneous sampled-data signal can be obtained by multiplying the spectrum of the sample pulse with

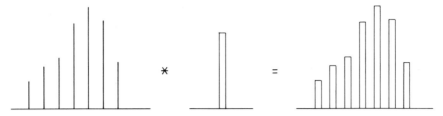

Figure 11.7 Instantaneous sampling.

the spectrum of the ideally sampled signal.

$$\mathscr{F}\left\{p(t) \otimes \left[x(t) \sum_{n=-\infty}^{\infty} \delta(t-nT)\right]\right\}$$

$$= P(f) \cdot \left\{X(f) \otimes \left[f_s \sum_{m=-\infty}^{\infty} \delta(f-mf_s)\right]\right\}$$

As shown in Fig. 11.8, the resulting spectrum is similar to the spectrum produced by ideal sampling. The only difference is the amplitude distortion introduced by the spectrum of the sampling pulse. This distortion is sometimes called the *aperture effect*. Notice that distortion is present in all the images, including the one at baseband. The distortion will be less severe for narrow sampling pulses. As the pulses become extremely narrow, instantaneous sampling begins to look just like ideal sampling, and distortion due to the aperture effect all but disappears.

Natural sampling

In natural sampling, each sample's amplitude follows the analog signal's amplitude throughout the sample's duration. As shown in Fig. 11.9, this is mathematically equivalent to multiplying the analog signal by a periodic train of rectangular pulses.

$$x_s(\cdot) = x(t) \cdot \left\{p(t) \otimes \left[\sum_{n=-\infty}^{\infty} \delta(t-nT)\right]\right\}$$

The spectrum of a naturally sampled signal is found by convolving the spectrum of the analog signal with the spectrum of the sampling pulse train.

$$\mathscr{F}[x_s(\cdot)] = X(f) \otimes \left[P(f)f_s \sum_{m=-\infty}^{\infty} \delta(f-mf_s)\right]$$

As shown in Fig. 11.10, the resulting spectrum will be similar to the

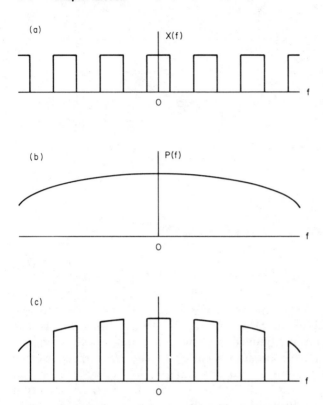

Figure 11.8 Spectrum (c) of an instantaneously sampled signal is equal to the spectrum (a) of an ideally sampled signal multiplied by the spectrum (b) of one sampling pulse.

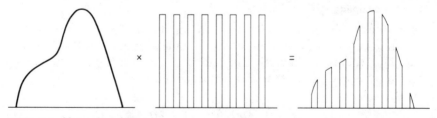

Figure 11.9 Natural sampling.

spectrum produced by instantaneous sampling. In instantaneous sampling, all frequencies of the sampled signal's spectrum are attenuated by the spectrum of the sampling pulse, while in natural sampling, each image of the basic spectrum will be attenuated by a factor which is equal to the value of the sampling pulse's spectrum at the center frequency of the image. In communications theory, natural sampling is called *shaped-top pulse amplitude modulation.*

Signal Processing and Simulation Issues   171

(a)
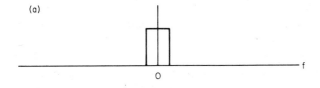

(b)
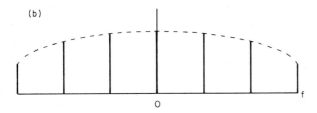

(c)
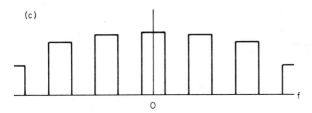

Figure 11.10 Spectrum (c) of a naturally sampled signal is equal to the spectrum (a) of the analog signal multiplied by the spectrum (b) of the sampling pulse train.

11.2 Discrete Fourier transform

The discrete Fourier transform (DFT) and its inverse are given by

$$X(m) = \sum_{n=0}^{N-1} x(n)e^{-j2\pi mnFT} \quad m = 0, 1, \ldots, N-1 \quad (11.1a)$$

$$= \sum_{n=0}^{N-1} x(n)\cos(2\pi mnFT) - j\sum_{n=0}^{N-1} x(n)\sin(2\pi mnFT) \quad (11.1b)$$

$$x(n) = \sum_{m=0}^{N-1} X(m)e^{j2\pi mnFT} \quad n = 0, 1, \ldots, N-1 \quad (11.2a)$$

$$= \sum_{m=0}^{N-1} X(m)\cos(2\pi mnFT) + j\sum_{m=0}^{N-1} X(m)\sin(2\pi mnFT) \quad (11.2b)$$

In designing a DFT for a particular application, values must be chosen for the parameters N, T, and F. N is the number of time sequence values $x(n)$ over which the DFT summation is performed to compute each frequency sequence value. It is also the total number of frequency sequence values $X(m)$ produced by the DFT. For convenience, the complete set of N consecutive time sequence values is referred to as the *input record*, and the complete set of N consecutive frequency sequence values is called the *output record*. T is the time interval between two consecutive samples of the input sequence, and F is the frequency interval between two consecutive samples of the output sequence. The selection of values for N, F, and T is subject to the following constraints, which are a consequence of the sampling theorem and the inherent properties of the DFT:

1. The inherent properties of the DFT require that $FNT = 1$.
2. The sampling theorem requires that $T \leq 1/(2f_H)$, where f_H is the highest significant frequency component in the continuous-time signal.
3. The record length in time is equal to NT or $1/F$.
4. Many "fast" DFT algorithms require that N be an integer power of 2.

A periodic function of time will have a discrete frequency spectrum, and a discrete-time function will have a spectrum that is periodic. Since the DFT relates a discrete-time function to a corresponding discrete-frequency function, this implies that both the time function and frequency function are periodic as well as discrete. This means that some care must be exercised in selecting DFT parameters and interpreting DFT results, but it does not mean that the DFT can be used only on periodic digital signals.

11.3 Simulation of White Gaussian Noise [Ref. 1]

The pseudorandom number generators commonly available in most high-level programming languages provide sequences of uniformly distributed numbers. The simulation of communication systems and processes usually requires gaussian-distributed numbers rather than uniformly distributed numbers. It is possible to take values from a sequence of independent, uniformly distributed numbers and from them generate a sequence of independent gaussian-distributed pseudorandom numbers using one of the two methods presented below.

Method A

1. Generate a pseudorandom value U_1 uniformly distributed on [0, 1).
2. Generate a second pseudorandom value U_2 which is uniform on [0, 1) and independent of U_1.

Signal Processing and Simulation Issues 173

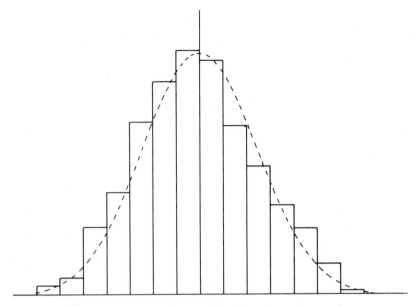

Figure 11.11 Histogram of a gaussian-distributed pseudorandom number generator.

3. Compute

$$G_1 = \cos(U_2)\sqrt{-2\sigma^2 \ln(U_1)}$$

$$G_2 = \sin(U_2)\sqrt{-2\sigma^2 \ln(U_1)}$$

The resulting G_1 and G_2 will be independent, zero-mean gaussian-distributed random variates each with variance σ^2. A histogram based on 10,000 samples of G_1 is shown in Fig. 11.11 along with a sketch of the corresponding theoretical probability density function.

Method B

1. Generate two pseudorandom values U_A and U_B which are independent and uniformly distributed on [0, 1).
2. Compute

$$U_1 = 1 - 2U_A$$

$$U_2 = 1 - 2U_B$$

3. Compute

$$S = U_1^2 + U_2^2$$

If $S \geq 1$, then go back to step 1; otherwise continue on to step 4.

4. Compute

$$G_1 = U_1 \sqrt{\frac{-2\sigma^2 \ln(S)}{S}}$$

$$G_2 = U_2 \sqrt{\frac{-2\sigma^2 \ln(S)}{S}}$$

The resulting G_1 and G_2 will be independent, zero-mean gaussian-distributed random variables each with variance σ^2.

11.4 Variance for Simulation of White Noise

In system performance specifications and theoretical analyses of communication systems, levels of white gaussian background noise are often specified in terms of the constant spectral density N_0. However, generation of gaussian-distributed pseudorandom numbers for simulating noise requires that a noise variance σ^2 be specified. In order to compare computer simulations against specifications and theoretical results, we need to establish a relationship between N_0 and σ^2.

As discussed in Sec. 5.3, the variance of a random noise process equals the process autocorrelation at lag zero:

$$\sigma^2 = R(0) \tag{11.3}$$

In the case of white noise (which is, recall, just an unrealizable but convenient idealization), $R(0)$ is infinite. However, in the analysis of communication system performance, we will usually be concerned with the noise power contained within some particular bandwidth of interest. Given a white noise density of N_0, the noise power in a bandwidth B is

$$P = \sigma^2 = N_0 B \tag{11.4}$$

To satisfy the uniform sampling theorem and prevent aliasing, the sampling rate must be chosen such that

$$R_s \geq \begin{cases} 2B & \text{for real sampling} \\ B & \text{for complex sampling} \end{cases} \tag{11.5}$$

Thus $\quad \sigma^2 = N_0 B \leq \begin{cases} \dfrac{N_0 R_s}{2} & \text{for real sampling} \\ N_0 R_s & \text{for complex sampling} \end{cases} \tag{11.6}$

If the noise is critically sampled to preserve whiteness, then the equalities in (11.5) and (11.6) will hold.

Example Consider a simulation of data transmission at 1200 bits/sec via a 4-ary FSK signal in additive white gaussian noise of spectral density N_0. The simulation program will generate 32 real samples during each symbol interval. Assume that the noise is critically sampled and that the signal tone has unity amplitude. Find the noise generator variance σ^2 which corresponds to $E_b/N_0 = 5$ dB, where E_b denotes energy per transmitted bit.

solution During each symbol interval, the energy in the signal tone is given by

$$E = \frac{A^2 T}{2} \tag{11.7}$$

where A is the signal tone amplitude and T is the symbol duration. Since each 4-ary symbol conveys 2 bits, the stated data rate of 1200 bits/sec corresponds to a symbol duration of $T = 1/600$. Thus, $E = 1/1200$. The energy per bit is one-half of the energy per symbol:

$$E_b = \frac{E}{2} = \frac{A^2 T}{4} = \frac{1}{2400} \tag{11.8}$$

The sampling rate is simply the symbol rate multiplied by the number of samples per symbol:

$$R_s = (32)(600) = 19{,}200 \tag{11.9}$$

From (11.6)
$$N_0 = \frac{2\sigma^2}{R_s} = \frac{\sigma^2}{9600} \tag{11.10}$$

Thus
$$\frac{E_b}{N_0} = \frac{1/2400}{\sigma^2/9600} = \frac{4}{\sigma^2} \tag{11.11}$$

The specified E_b/N_0 of 5 dB is easily converted into a pure ratio using

$$\frac{E_b}{N_0} = 10^{[(E_b/N_0)\,\text{dB}/10]} = 10^{0.5} \tag{11.12}$$

Substituting this into (11.11) we obtain

$$\sigma^2 = 4 \times 10^{-0.5}$$
$$= 1.2649 \tag{11.13}$$

11.5 Simulation of Noise in the Frequency Domain

In many signal processing schemes, the input signal is digitized and then a discrete Fourier transform is computed. All subsequent processing is then performed in the frequency domain using the DFT outputs. Two alternative approaches exist for simulating such a system. In the straightforward approach we would begin by generating a sequence of idealized signal samples and then adding samples from an appropriate noise process. This would be followed by computation of the DFT for a segment of the resulting signal-to-noise input sequence.

A much faster-running simulation is possible by using an approach based on direct synthesis of the DFT outputs. The DFT outputs for a noise-free input signal can be synthesized directly from theoretical considerations. To these DFT outputs we can then add samples from the appropriate noise process *in the frequency domain* which corresponds to the time domain noise process to be simulated.

Example Consider noncoherently detected 4-ary FSK in white gaussian noise. Assume that a sinusoidal tone will be transmitted on one of four frequencies—f_0, f_1, f_2, or f_3. These frequencies correspond to four of the FFT output frequencies. (There may be many more FFT output frequencies, but they are not of interest as they fall between valid tone frequencies.) The time domain approach entails the following steps:

1. Generate a digitized sine wave at one of the four tone frequencies.
2. Add samples of white gaussian noise to each of the sinusoid samples.
3. Using a discrete Fourier transform, compute the magnitude spectrum at frequencies f_0, f_1, f_2, and f_3 for a segment of the signal-plus-noise input sequence.
4. Examine the magnitudes at each of the four tone frequencies to determine which tone was transmitted.

For an N-point input sequence, evaluation of the DFT at the four frequencies of interest involves $8N$ real multiplications and additions. Furthermore, N pseudorandom gaussian noise samples must be generated and added to the signal.

The frequency domain approach entails the following steps:

1. Directly synthesize the spectrum for the sinusoidal tone. This will consist of an appropriate complex constant at the frequency corresponding to the transmitted tone and zeros at all other frequencies. (In this example we assume that the orthogonality between the four tone frequencies is preserved by any processing which occurs prior to the digitizer.)
2. As shown in Sec. 8.3, the Fourier coefficients of a discrete-time bandpass gaussian process are themselves pairs of gaussian random variables, and the magnitude (envelope) value at each frequency will be a Rayleigh-distributed random variable. Therefore, to simulate the noise, just generate three samples from a Rayleigh random process; and use these values to simulate the noise spectrum at the three frequencies where the signal spectrum contains zeros.
3. Generate a quadrature pair of independent gaussian-distributed random values, and add them to the constant representing the spectrum of the sine tone determined in step 1. Form the envelope of the resulting quadrature pair, that is,

$$E = \sqrt{(A_R + G_1)^2 + (A_I + G_2)^2}$$

where $A_R + jA_I$ = DFT value at the tone frequency
G_1, G_2 = quadrature pair of gaussian variates

This approach does not involve evaluation of a DFT. Direct synthesis of the

DFT outputs entails generation of three pseudorandom Rayleigh-distributed noise samples and one pair of independent gaussian noise samples.

11.6 References

1. D. E. Knuth: *The Art of Computer Programming: Vol. 2, Seminumerical Algorithms*, 2d ed., Addison-Wesley, Reading, Mass., 1981.
2. B. Rorabaugh: *Signal Processing Design Techniques*, TAB Professional and Reference Books, Blue Ridge Summit, Penn., 1986.

Chapter

12

Continuous Modulation

12.1 Amplitude Modulation [Refs. 1 and 4]

An amplitude modulated (AM) signal $s(t)$ can be mathematically represented as:

$$s(t) = A[1 + kx(t)] \cos(2\pi f_c t + \theta) \quad (12.1)$$

or $$s(t) = A[1 + \mu m(t)] \cos(2\pi f_c t + \theta) \quad (12.2)$$

where $x(t)$ = message signal
 k = amplitude sensitivity
 $m(t) = x(t)/A_m$ with A_m chosen such that $\max|m(t)| = 1$
 $\mu = kA_m$ is the modulation index
 f_c = carrier frequency
 θ = an arbitrary (but constant) phase shift
 A = carrier amplitude

It is assumed that the carrier frequency f_c is large compared to the bandwidth of the message signal $x(t)$. In many types of analysis where the carrier phase θ is not important, it is simply assumed to be zero.

An unmodulated carrier, a message signal, and the corresponding AM signal are shown in Fig. 12.1. The dotted trace in Fig. 12.1c represents the *envelope* of $s(t)$. Note that the upper trace of the envelope has the same shape as the message signal, but the former is vertically offset from the latter by an amount equal to A. The lower envelope trace is simply the mirror image of the upper trace reflected across the time axis. When $\mu \leq 1$, the message signal $m(t)$ and the envelope of $s(t)$ will match in shape. However, when $\mu > 1$, a condition known as *overmodulation* exists and the envelope of $s(t)$ will

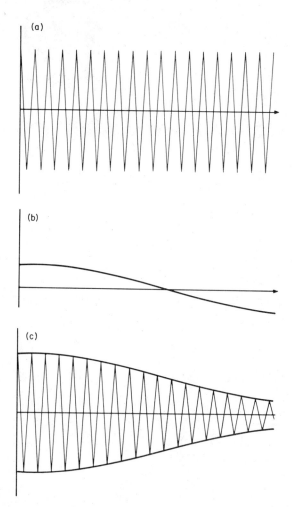

Figure 12.1 Components of an AM signal. (*a*) Carrier $\cos(2\pi f_c t + \theta)$. (*b*) Message signal $m(t)$. (*c*) Modulated signal $s(t)$.

cross the time axis as shown in Fig. 12.2. This is an undesirable condition because an envelope detector will produce a signal equal to the envelope's absolute value which is depicted by the solid line.

Spectrum of AM signals

The AM signal defined by Eq. (12.2) will have a spectrum $S(f)$ which is given by

$$S(f) = \frac{A}{2}[\delta(f - f_c) + \delta(f + f_c)] + \frac{\mu A}{2}[M(f - f_c) + M(f + f_c)] \quad (12.3)$$

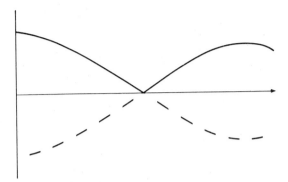

Figure 12.2 Distorted envelope due to overmodulation.

where $M(f)$ is the Fourier transform of $m(t)$. Equation (12.3) follows directly from the Fourier transform property 8 given in Table 6.2. Figure 12.3 depicts the relationship between $S(f)$ and $M(f)$. The shaded portions in Fig. 12.3b are the *upper sidebands* (USB) of the AM signal which correspond to the shaded portion of the message signal in part a of the figure. The unshaded portions represent the *lower sidebands* (LSB).

Power in an AM signal

If the signal $s(t)$ defined by Eq. (12.2) represents a voltage or current waveform, the total power P_T delivered to a 1-Ω resistor is given by

$$P_T = \frac{A^2}{2} + \langle m^2(t) \rangle \frac{A^2\mu^2}{2} \qquad (12.4)$$

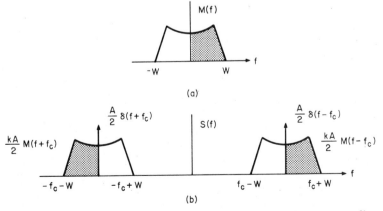

Figure 12.3 Magnitude spectra of a message signal (a) and the corresponding AM signal (b).

where $\langle m^2(t)\rangle$ denotes the time average of $m^2(t)$. The first term in (12.4) represents the power in the carrier component, and the second term represents the power in the two sidebands. Since the useful information is contained solely in these sidebands, it is advantageous to maximize the fraction of total transmitted power which they contain. The absolute value of $m(t)$ never exceeds unity; therefore, it follows that $\langle m^2(t)\rangle \le 1$.

Square wave modulation

For a binary message signal which switches between plus and minus 1, $m^2(t)$ will equal 1, and the sideband power P_T will be maximized:

$$P_T = \frac{A^2}{2} + \frac{A^2\mu^2}{2} \qquad (12.5)$$

The sideband power expressed as a fraction of the total power is given by

$$\frac{P_{SB}}{P_T} = \frac{A^2\mu^2/2}{A^2/2 + A^2\mu^2/2} = \frac{\mu^2}{1+\mu^2} \qquad (12.6)$$

At 100 percent modulation ($\mu = 1$), only one-half of the total power is contained in the sidebands. For lower values of $\mu < 1$, the fraction of total power allocated to the information-bearing sidebands will be even smaller as shown by the plot of sideband power versus modulation index contained in Fig. 12.4.

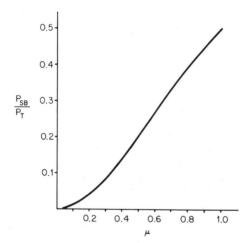

Figure 12.4 Fraction of power allocated to sidebands versus modulation index for an AM signal modulated by a binary square wave.

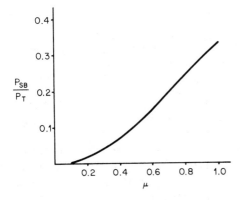

Figure 12.5 Fraction of power allocated to sidebands versus modulation index for an AM signal modulated by a sinusoid.

Sinusoidal modulation

For sinusoidal $m(t)$, we find $\langle m^2(t) \rangle = 1/2$ and the fraction of total power in the sidebands equals $\mu^2/(2+\mu^2)$. At 100 percent modulation, only one-third of the total power is contained in the sidebands. A plot of sideband power versus modulation index for sinusoidal modulation is shown in Fig. 12.5.

12.2 Modulators for AM

Square-law modulator [Refs. 1, 2, and 3]

A conceptual model of a *square-law modulator* for generation of AM signals is depicted in Fig. 12.6. The definition of a square-law device indicates that

$$s_2(t) = as_1(t) + bs_1^2(t) \tag{12.7}$$

where the values of a and b are determined by the characteristics of the square-law device. Since

$$s_1(t) = A\cos(2\pi f_c t) + m(t) \tag{12.8}$$

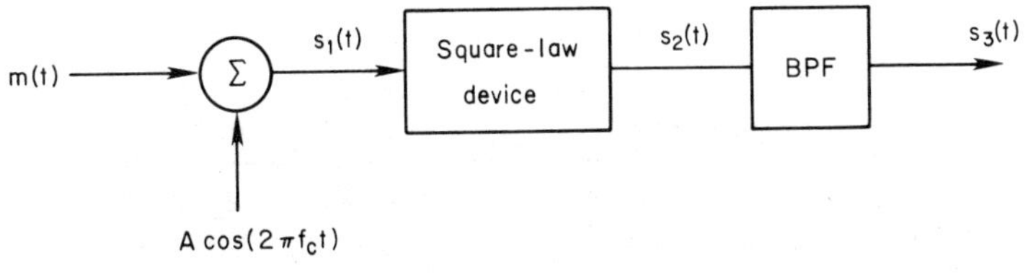

Figure 12.6 Square-law modulator.

simple substitution into (12.7) reveals that

$$s_2(t) = aA\left[1 + \frac{2b}{a}m(t)\right]\cos(2\pi f_c t) + am(t) + bm^2(t) + bA^2\cos^2(2\pi f_c t) \quad (12.9)$$

The spectrum of $s_2(t)$ is depicted in Fig. 12.7. The desired AM signal is represented by the first term in (12.10) and occupies frequency bands $-(f_c + W)$ to $-(f_c - W)$ and from $(f_c - W)$ to $(f_c + W)$ (that is, the components labeled 2, 3, 7, and 8). The second and third terms are baseband components occupying frequencies from $-W$ to W (labeled 5) and from $-2W$ to $2W$ (labeled 4), respectively. Using trigonometric identity given by Eq. (2.26), the fourth term can be rewritten as:

$$bA^2\cos^2(2\pi f_c t) = \frac{bA^2}{2} + \frac{bA^2}{2}\cos(4\pi f_c t) \quad (12.10)$$

The spectrum components corresponding to (12.10) will consist of impulses at $f = 0$ (labeled 6) and at $f = \pm 2f_c$ (labeled 1 and 9). The characteristics of the bandpass filter are selected so that the AM signal lies within the passband, while the baseband and double-frequency components lie within the stop band, thus yielding

$$s_3(t) = GaA\left[1 + \frac{2b}{a}m(t)\right]\cos(2\pi f_c t) \quad (12.11)$$

where G is a constant gain factor determined by the characteristics of the bandpass filter. It should be noted that if W is greater than $f_c/3$, component 4 will overlap with components 2 and 7, making the extraction of s_3 impossible.

A circuit for implementing a square-law modulator is shown in Fig. 12.8. The carrier signal and modulating signal are coupled into the base-emitter circuit by transformers T_1 and T_2, respectively. The

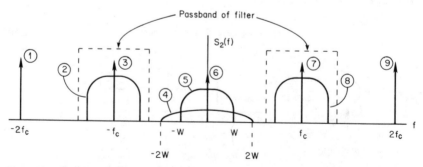

Figure 12.7 Spectrum of square-law modulator output prior to bandpass filtering.

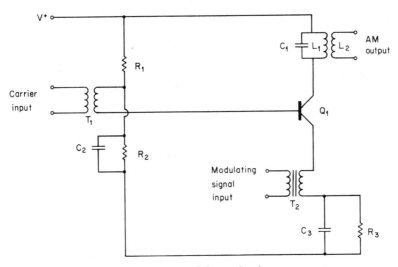

Figure 12.8 A practical square-law modulator circuit.

resistors R_1, R_2, and R_3 are selected to bias Q_1 for nonlinear operation. Values for C_1 and L_1 are selected so that the collector circuit will resonate at the carrier frequency, thus acting as a bandpass filter to select the desired AM components (i.e., the components labeled 2, 3, 7, and 8 in Fig. 12.7). The selected bandwidth will depend upon the Q of the C_1–L_1 resonator.

Switching modulator [Refs. 1 and 2]

A conceptual model of a *switching modulator* for generation of AM signals is depicted in Fig. 12.9. It is assumed that $v_1(t)$ is large compared to the bias on the diode, thus causing the diode to switch between conduction and cutoff as $v_1(t)$ swings above and below the

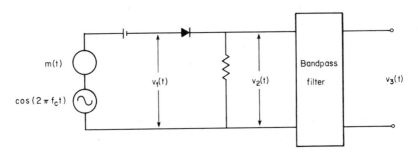

Figure 12.9 Switching modulator.

bias level. The input voltage is given by

$$v_1(t) = A\cos(2\pi f_c t) + m(t) \qquad (12.12)$$

If the magnitude of $m(t)$ is much smaller than A, the conduction of cutoff state of the diode will be determined primarily by the polarity of the carrier, and the voltage $v_2(t)$ can be approximated by

$$v_2(t) \cong \begin{cases} v_1(t) & \cos(2\pi f_c t) > 0 \\ 0 & \cos(2\pi f_c t) < 0 \end{cases} \qquad (12.13)$$

Equivalently, we can eliminate the brace and conditional outputs in (12.13) by introducing a periodic pulse train $p(t)$ which enables us to write

$$v_2(t) \cong p(t)[A\cos(2\pi f_c t) + m(t)] \qquad (12.14)$$

As shown in Fig. 12.10, $p(t) = 1$ for the values of t which make $\cos(2\pi f_c t)$ positive, and $p(t) = 0$ for the values of t which make $\cos(2\pi f_c t)$ negative. Expanded as a Fourier series, the pulse train can be expressed as:

$$p(t) = \frac{1}{2} + \frac{2}{\pi}\sum_{n=1}^{\infty}\frac{(-1)^{n-1}}{2n-1}\cos[2(2n-1)\pi f_c t] \qquad (12.15)$$

Substituting (12.15) into (12.14) and vigorously applying some trigonometric identities, we can obtain

$$\begin{aligned}v_2(t) \cong{} & \frac{A}{2}\cos(2\pi f_c t)\left[1 + \frac{4}{\pi A}m(t)\right] + \frac{Am(t)}{2} \\ & + \frac{A}{\pi} + \frac{A}{\pi}\sum_{n=1}^{\infty}\frac{2(-1)^{n-1}}{4n^2-1}\cos(4\pi n f_c t) \\ & + \frac{2m(t)}{\pi}\sum_{n=2}^{\infty}\frac{(-1)^{n-1}}{2n-1}\cos[2(2n-1)\pi f_c t]\end{aligned} \qquad (12.16)$$

The spectrum of $v_2(t)$ is depicted in Fig. 12.11. The desired AM signal

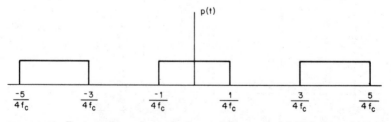

Figure 12.10 Rectangular pulse train for modeling a diode's switching function.

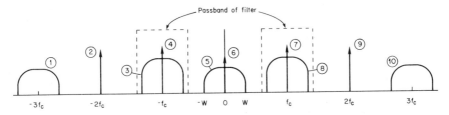

Figure 12.11 Spectrum of signal components present in a switching modulator.

is represented by the first term in (12.16) and occupies frequency bands from $-(f_c + W)$ to $(f_c - W)$ and from $(f_c - W)$ to $(f_c + W)$ (i.e., the components labeled 3, 4, 7, and 8). The second term is a baseband component occupying frequencies from $-W$ to W (labeled 5), and the third term is a dc component represented by an impulse at $f = 0$ (labeled 6). The summation comprising the fourth term represents a series of impulses at frequencies of $\pm k f_c$ where $k = 2, 4, 6, \ldots$. (Figure 12.11 is only large enough to show the components at $\pm 2 f_c$ labeled 2 and 9.) The fifth term represents a series of translated images of the modulating signal centered at frequencies of $\pm k f_c$ where $k = 3, 5, 7, \ldots$.

Just as for the switching modulator, the characteristics of the bandpass filter must be selected so that the AM signal lies within the passband, while the baseband and harmonic components lie within the stop band, thus yielding

$$v_3(t) \cong \frac{A}{2}\left[1 + \frac{4}{\pi A} m(t)\right] \cos(2\pi f_c t) \qquad (12.17)$$

Note that if W is greater than $f_c/2$, components 3 and 8 will overlap with component 5 making the extraction of v_3 via bandpass filtering impossible. Assuming $\max|m(t)| = 1$, the modulation index is $\mu = 4/(\pi A)$.

Switching modulators are sometimes called *chopper modulators* or *diode modulators*. Switching modulators are most often implemented as balanced modulators.

12.3 Measures of Demodulator Performance

Older texts use *demodulator gain* as a basis for comparison of various demodulation schemes. This demodulator gain is simply the output SNR of the demodulator divided by the input SNR:

$$G_{\text{demod}} = \frac{SNR_{\text{out}}}{SNR_{\text{in}}} \qquad (12.18)$$

As demonstrated in Sec. 12.4, both signal-to-noise ratios are computed without regard to necessary bandwidths at the input and output. Texts (such as Refs. 6 and 7) which use demodulator gain, generally use the mean-square approach to compute the noise powers for the SNRs. Only the message components of the input and output signals are used for computing the signal powers.

A more modern approach for comparison of demodulation schemes involves a *figure of merit*, usually denoted by γ and defined by

$$\gamma = \frac{(SNR)_0}{(SNR)_c} = \frac{S_{out}/N_{out}}{S_{in}/N_m} \qquad (12.19)$$

where S_{out} = average power of the message signal at the demodulator output
N_{out} = average noise power at the demodulator output
S_{in} = average power of the modulated message signal at the demodulator input
N_m = average power of noise *in the message bandwidth* at the demodulator input

Texts, such as Refs. 1 and 4, which use figure of merit as a basis for comparison, use the bandwidth and spectral density to compute N_{out} and N_m. There is one apparent inconsistency in the determination of S_{out} and S_{in} which should be discussed. In computing S_{in}, the power due to a carrier component is included if present in the input as it is for conventional AM. However, constant components in the output which are due to demodulated carrier components are neglected in the computation of S_{out}. The inclusion of carrier power in S_{in} tends to lower the figure of merit and thus treats carrier power as a cost. The exclusion of carrier components from S_{out} is consistent with this viewpoint, since these components convey no information about the message. Inclusion of these components would tend to increase the figure of merit and thus inappropriately treat power allocated to the carrier as a benefit.

12.4 Envelope Demodulation

An AM signal can be demodulated using an envelope demodulator consisting of a rectifier followed by a lowpass filter as shown in Fig. 12.12. The exact analysis of an envelope demodulator is quite difficult, and it is convenient to avoid a strict mathematical development in favor of the following intuitive argument regarding the demodulating action of the lowpass filter. Assume that the lowpass filter is an R-C network having a time constant which is short relative to changes in

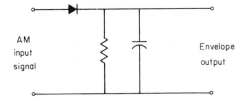

Figure 12.12 Envelope demodulator for AM signals.

the modulating signal $m(t)$ but long relative to the period $1/f_c$ of the carrier. Thus the decay of the voltage across the capacitor is too slow for any significant change from cycle to cycle of the carrier, yet fast enough to faithfully reproduce the envelope variations corresponding to $m(t)$. Given an input signal of the form:

$$x(t) = A[1 + \mu m(t)] \cos(2\pi f_c t) + n(t) \qquad (12.20a)$$

where $n(t) = n_c(t) \cos(2\pi f_c t) - n_s(t) \sin(2\pi f_c t)$ \qquad (12.20b)

an envelope demodulator will exhibit the following characteristics provided that the input SNR is sufficiently high.

Output signal: $\quad A[1 + \mu m(t)] + n_c(t)$

Demodulation gain: $\quad 2$

Figure of merit: $\quad \dfrac{\mu^2 P_m}{1 + \mu^2 P_m}$

where P_m is the average power in the message signal $m(t)$.

In cases where the input SNR is low, the figure of merit is approximated by

$$\gamma = \frac{A^2 \mu^2 P_m + A^2 \mu^4 P_m^2}{2.2 W N_0} \qquad (12.21)$$

Demodulation gain

The following derivation of demodulation gain for the envelope demodulator is presented in detail because it contains several examples of the kind of assumptions and mathematical manipulations that are often involved in the analysis of amplitude-modulated signals.

Consider a noisy input signal, $x(t)$, given by

$$x(t) = s(t) + n(t) \tag{12.22a}$$
$$= A[1 + \mu m(t)] \cos(2\pi f_c t) + n_c(t) \cos(2\pi f_c t)$$
$$+ n_s(t) \sin(2\pi f_c t) \tag{12.22b}$$
$$= B(t) \cos[2\pi f_c t + \phi(t)] \tag{12.22c}$$

where
$$B(t) = \{[A + A\mu m(t) + n_c(t)]^2 + n_s^2(t)\}^{1/2} \tag{12.22d}$$
$$= [A^2 + 2A^2\mu m(t) + 2An_c(t) + A^2\mu^2 m^2(t)$$
$$+ 2A\mu m(t)n_c(t) + n_c^2(t) + n_s^2(t)]^{1/2} \tag{12.22e}$$

$$\phi(t) = \tan^{-1}\left[\frac{n_s(t)}{A + a\mu m(t) + n_c(t)}\right] \tag{12.22f}$$

An ideal envelope detector would produce an output equal to $B(t)$ which is the envelope of $x(t)$. Exact analysis of the envelope demodulator's performance is quite involved, and at this point it is convenient to introduce a simplifying assumption. For cases where the signal is much stronger than the noise, the quadrature noise component $n_s(t)$ can be assumed to be zero. Although $n_s(t)$ and $n_c(t)$ will have equal powers, the inphase component $n_c(t)$ must be retained since it is multiplied by large signal terms in Eq. (12.22e). Under this assumption, $B(t)$ becomes

$$B(t) = A + A\mu m(t) + n_c(t) \tag{12.23}$$

The useful, information-bearing component $s_0(t)$ of the output is equal to $A\mu m(t)$, with a power given by

$$P_{s_o} = \langle s_0^2(t) \rangle = A^2\mu^2 \langle m^2(t) \rangle \tag{12.24}$$

The output noise power is given by

$$P_{n_o} = \langle n_c^2(t) \rangle \tag{12.25}$$

Thus the output SNR is given by

$$SNR_{out} = \frac{A^2\mu^2 \langle m^2(t) \rangle}{\langle n_c^2(t) \rangle} \tag{12.26}$$

The signal power at the input is given by

$$P_{s_i} = \langle s^2(t) \rangle = \langle A^2\mu^2 m^2(t) \cos^2[2\pi f_c t + \phi(t)] \rangle \tag{12.27}$$

Applying the trigonometric identity of Eq. (2.26) we can rewrite Eq. (12.27) as:

$$P_{s_i} = \left\langle \frac{A^2\mu^2 m^2(t)}{2} \{1 + \cos[4\pi f_c t + 2\phi(t)]\} \right\rangle \qquad (12.28)$$

Assuming that $m^2(t)$ and the cosine term are uncorrelated, Eq. (12.28) reduces to $P_{s_i} = A^2\mu^2 \langle m^2(t) \rangle / 2$. Thus the input SNR is given by

$$SNR_{\text{in}} = \frac{A^2\mu^2 \langle m^2(t) \rangle / 2}{\langle n_c^2(t) \rangle} \qquad (12.29)$$

The demodulation gain is finally obtained as:

$$\begin{aligned} G_{\text{demod}} &= \frac{SNR_{\text{out}}}{SNR_{\text{in}}} \\ &= \left[\frac{A^2\mu^2 \langle m^2(t) \rangle}{\langle n_c^2(t) \rangle}\right]\left[\frac{\langle n_c^2(t) \rangle}{A^2\mu^2 \langle m^2(t) \rangle / 2}\right] \\ &= 2 = 3 \text{ dB} \end{aligned} \qquad (12.30)$$

For low input SNR, the assumption made to obtain Eq. (12.23) is not valid; and it can be shown [Ref. 8] that G_{demod} will be less than indicated by Eq. (12.30).

Figure of merit

For clarity of presentation and for the convenience of the reader, the following derivation of figure of merit for the envelope demodulator is presented in its entirety, even though some material from the derivation of demodulator gain is repeated.

Consider a noisy signal $x(t)$ given by

$$x(t) = s(t) + n(t) \qquad (12.31a)$$

where $s(t) = A[1 + \mu m(t)] \cos(2\pi f_c t) \qquad (12.31b)$

$$n(t) = n_c(t) \cos(2\pi f_c t) + n_s(t) \sin(2\pi f_c t) \qquad (12.31c)$$

The average power in the input signal is given by

$$S_i = \langle A^2 \cos^2(2\pi f_c t) \rangle (1 + \mu^2 P_m) \qquad (12.32a)$$

$$= \left\langle \frac{A^2}{2}[1 + \cos(4\pi f_c t)] \right\rangle (1 + \mu^2 P_m) \qquad (12.32b)$$

$$= \left[\left\langle \frac{A^2}{2} \right\rangle + \left\langle \frac{A^2}{2} \cos(4\pi f_c t) \right\rangle\right](1 + \mu^2 P_m) \qquad (12.32c)$$

$$= \left[\frac{A^2}{2} + 0\right][1 + \mu^2 P_m] \qquad (12.32d)$$

$$= \frac{A^2[1 + \mu^2 P_m]}{2} \qquad (12.32e)$$

where P_m is the average power in the message signal. Assuming that the noise is AWGN with a two-sided power spectral density of $N_0/2$ and that the message bandwidth is W, then the noise term N_m is equal to WN_0. Thus

$$(SNR)_c = \frac{A^2(1+\mu^2 P_m)}{2WN_0} \qquad (12.33)$$

The input signal given by Eq. (12.31a) can be rewritten as:

$$x(t) = B(t)\cos[2\pi f_c t + \phi(t)] \qquad (12.34a)$$

where $B(t) = \{[A + A\mu m(t) + n_c(t)]^2 + n_s^2(t)\}^{1/2}$

$$= [A^2 + 2A^2\mu m(t) + 2An_c(t) + A^2\mu^2 m^2(t)$$
$$+ 2A\mu m(t)n_c(t) + n_c^2(t) + n_s^2(t)]^{1/2} \qquad (12.34b)$$

$$\phi(t) = \tan^{-1}\left[\frac{n_s(t)}{A + a\mu m(t) + n_c(t)}\right] \qquad (12.34c)$$

An ideal envelope detector would produce an output equal to $B(t)$ which is the envelope of $x(t)$. Exact analysis of the envelope demodulator's performance is quite involved, and at this point it is convenient to introduce a simplifying assumption. For cases where the signal is much stronger than the noise, the quadrature noise component $n_s(t)$ can be assumed to be zero. Although $n_s(t)$ and $n_c(t)$ will have equal powers, the inphase component $n_c(t)$ must be retained since it is multiplied by large signal terms in Eq. (12.34c). Under this assumption, $B(t)$ becomes

$$B(t) = A + A\mu m(t) + n_c(t) \qquad (12.35)$$

The message signal at the demodulator output is equal to $A\mu m(t)$ which has a power given by

$$S_{\text{out}} = \langle A^2\mu^2\rangle P_m$$
$$= A^2\mu^2 P_m \qquad (12.36)$$

As noted in Sec. 12.3, the constant output component [first term on the RHS of Eq. (12.35)] due to demodulated carrier is not considered as part of the message signal. Given the psd shown in Fig. 12.13a for $n(t)$, the psd for $n_c(t)$ will be as shown in Fig. 12.13b. Since the noise component of the output is $n_c(t)$, the output noise power is equal to

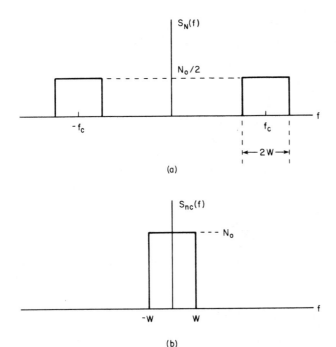

Figure 12.13 Spectral densities of noise in envelope demodulation. (a) psd for $n(t)$. (b) psd for $n_c(t)$.

the area under the psd of Fig. 12.13b, or just $2WN_0$. Thus

$$(SNR)_0 = \frac{A^2\mu^2 P_m}{2WN_0} \tag{12.37}$$

and

$$\gamma = \frac{A^2\mu^2 P_m/2WN_0}{A^2(1+\mu^2 P_m)/2WN_0} = \frac{\mu^2 P_m}{1+\mu^2 P_m} \tag{12.38}$$

12.5 Square-Law Detector [Ref. 1]

A square-law detector used for AM demodulation is shown in Fig. 12.14. The AM input signal is given by

$$x(t) = A[1+\mu m(t)]\cos(2\pi f_c t) \tag{12.39}$$

An appropriate selection of the lowpass filter will yield an output $y(t)$ which is given by

$$y(t) = K[1+2\mu m(t)+\mu^2 m^2(t)] \tag{12.40}$$

where K is the (assumed constant) passband gain of the lowpass

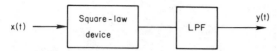

Figure 12.14 Square-law detector.

filter. The first term is just a dc offset. The second term is the desired output which is proportional to the original message signal $m(t)$. The third term represents unwanted distortion which falls within the passband of the lowpass filter. The ratio of desired output of unwanted distortion, or signal-to-distortion ratio (SDR), is given by

$$SDR = \frac{2\mu m(t)}{\mu^2 m^2(t)} = \frac{2}{\mu m(t)} \qquad (12.41)$$

To maximize the SDR, the magnitude of the denominator in Eq. (12.41) must be minimized.

12.6 Coherent Demodulation of AM Signals

Coherent demodulation of an AM signal can be performed as indicated in the block diagram of Fig. 12.15. Given an input signal of the form:

$$x(t) = A[1 + \mu m(t)] \cos(2\pi f_c t) + n(t) \qquad (12.42a)$$

where $n(t) = n_c(t) \cos(2\pi f_c t) - n_s(t) \sin(2\pi f_c t)$ $\qquad (12.42b)$

a coherent demodulator will exhibit the following characteristics:

Output signal: $\quad \dfrac{A[1 + \mu m(t)] + n_c(t)}{2}$

Demodulation gain: $\quad 2$

Figure of merit: $\quad \dfrac{\mu^2 P_m}{1 + \mu^2 P_m}$

where P_m is the average power in the message signal $m(t)$.

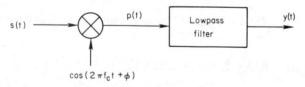

Figure 12.15 Coherent demodulator for AM signals.

Note that the demodulation gain is always 3 dB regardless of the input SNR. The difficulty of providing a reference signal that is phase- and frequency-locked to the carrier can be considerable; but in cases of low SNR, coherent demodulation can provide performance that is superior to envelope detection. Coherent demodulation is also called *synchronous detection* or *homodyne detection*.

12.7 Double-Sideband Suppressed-Carrier (DSBSC) Modulation

In a modified form of AM called *double-sideband suppressed-carrier* (DSBSC) modulation, the carrier component is totally eliminated from the transmitted signal. A DSBSC signal can be mathematically represented as:

$$s(t) = Am(t)\cos(2\pi f_c t) \qquad (12.43)$$

As shown in Fig. 12.16, a DSBSC signal exhibits a phase reversal each time the polarity of $m(t)$ changes. This is similar to overmodulation in conventional AM.

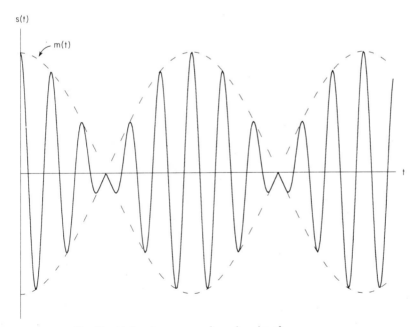

Figure 12.16 Double-sideband, suppressed-carrier signal.

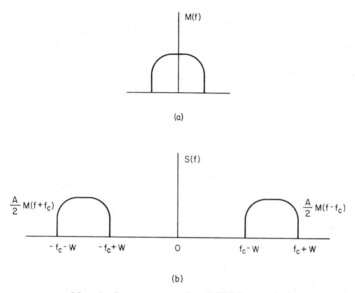

Figure 12.17 Magnitude spectrum of a DSBSC signal. Spectrum of baseband modulating signal is shown in (a). Spectrum of modulated signal is shown in (b).

Spectrum of DSBSC signals

The DSBSC signal defined by Eq. (12.43) will have a spectrum given by

$$S(f) = \frac{A}{2}[M(f - f_c) + M(f + f_c)] \qquad (12.44)$$

where $M(f)$ represents the Fourier transform of $m(t)$. This spectrum is depicted in Fig. 12.17. Except for the absence of impulses at $\pm f_c$, the DSBSC spectrum is similar to the AM spectrum shown in Fig. 12.3.

Generation of DSBSC using a balanced modulator

A DSBSC signal can be generated with the balanced modulator shown in Fig. 12.18. The boxes marked "amplitude modulator" are conventional amplitude modulators providing outputs $s_1(t)$ and $s_2(t)$ given by

$$s_1(t) = A[1 + \mu m(t)] \cos(2\pi f_c t) \qquad (12.45)$$

$$s_2(t) = A[1 - \mu m(t)] \cos(2\pi f_c t) \qquad (12.46)$$

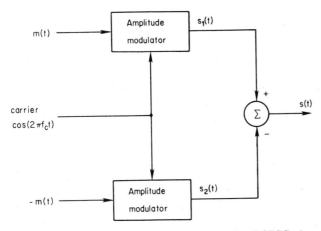

Figure 12.18 Balanced modulator for generating DSBSC signals.

The output signal $s(t)$ is given by

$$s(t) = s_1(t) - s_2(t) = 2\mu m(t) \cos(2\pi f_c t) \qquad (12.47)$$

which is just the DSESC signal defined by Eq. (12.43) with $A = 2\mu$.

Demodulation of DSBSC signals

The coherent demodulator shown in Fig. 12.15 can be used to demodulate DSBSC signals. For a DSBSC input signal given by

$$s(t) = Am(t) \cos(2\pi f_c t + \theta) \qquad (12.48)$$

and a local reference signal

$$r(t) = B \cos(2\pi f_c t + \theta) \qquad (12.49)$$

the multiplier output $p(t)$ will be

$$\begin{aligned} p(t) &= Am(t) \cos(2\pi f_c t + \theta) B \cos(2\pi f_c t + \theta) \\ &= ABm(t) \cos^2(2\pi f_c t + \theta) \\ &= \frac{AB}{2} m(t) + \frac{AB}{2} m(t) \cos 2(2\pi f_c t + \theta) \end{aligned} \qquad (12.50)$$

With an appropriately specified lowpass filter, the output $y(t)$ is given by

$$y(t) = \frac{AB}{2} m(t) \qquad (12.51)$$

The major problem in coherent demodulation of DSBSC signals is the difficulty in obtaining a reference signal $r(t)$ which is phase-locked to the carrier. If the local reference is not exactly in phase with the received carrier, the output will be different from that indicated by Eq. (12.51). Consider a local reference signal with a constant phase ϕ which may differ from the phase θ of the carrier

$$r(t) = B\cos(2\pi f_c t + \phi) \qquad (12.52)$$

The product $p(t)$ will then be

$$p(t) = \frac{AB}{2} m(t) \cos(\theta - \phi) + \frac{AB}{2} m(t) \cos(4\pi f_c t + \phi + \theta) \qquad (12.53)$$

After lowpass filtering, this will yield an output of

$$y(t) = \frac{AB}{2} m(t) \cos(\theta - \phi) \qquad (12.54)$$

Equations (12.51) and (12.54) differ by a factor equal to the cosine of the phase error between the carrier and local reference.

12.8 Single sideband

Consider the DSBSC signal defined by

$$s(t) = Am(t) \cos(2\pi f_c t) \qquad (12.55)$$

where $m(t)$ is the message signal having a spectrum $M(f)$ as depicted in Fig. 12.3a. We can make the following observations:

1. If $m(t)$ is real-valued, then $M(f)$ will be symmetric about $f = 0$.
2. As discussed in Secs. 12.1 and 12.7, the spectra of DSBSC and conventional AM signals will each contain components corresponding to $M(f)$ which are centered at $\pm f_c$ as shown in Figs. 12.3b and 12.17b. As a consequence of 1, each of these components will exhibit local symmetry about $f = \pm f_c$.
3. Given that the DSBSC and conventional AM signals under discussion are real-valued, then the spectra $S_{\text{DSBSC}}(f)$ and $S_{\text{AM}}(f)$ will each also exhibit symmetry about $f = 0$. (The Fourier transform of a real-valued function will have a real part with even symmetry plus an imaginary part with odd symmetry.)
4. Half of each symmetric component can be removed without loss of information since knowledge of the symmetry allows the missing half to be uniquely reconstructed from the remaining half. If both upper sidebands are removed as in Fig. 12.19a or if both lower

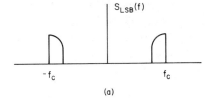

(a)

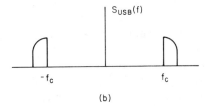

(b)

Figure 12.19 Magnitude spectra for (a) LSB single-sideband signal and (b) USB single-sideband signal.

sidebands are removed as in Fig. 12.19b, the overall symmetry of the remaining spectrum will not be disturbed. Thus the corresponding signal will remain real-valued.

5. The mathematical representations of the time domain signals corresponding to $S_{\text{USB}}(f)$ and $S_{\text{LSB}}(f)$ are given by

$$s_{\text{USB}}(t) = \frac{Am(t)}{2}\cos(2\pi f_c t) - \frac{A\hat{m}(t)}{2}\sin(2\pi f_c t) \quad (12.56)$$

$$s_{\text{LSB}}(t) = \frac{Am(t)}{2}\cos(2\pi f_c t) + \frac{A\hat{m}(t)}{2}\sin(2\pi f_c t) \quad (12.57)$$

where $\hat{m}(t)$ denotes the Hilbert transform of $m(t)$.

6. SSB has the obvious advantage of requiring only half the bandwidth of the corresponding DSBSC or AM signal.

The single-sideband (SSB) signals defined by Eqs. (12.56) and (12.57) are the classical forms presented in most communications textbooks. There is a more general representation which encompasses several variants of SSB signals which may be encountered in practice. This general representation is given by

$$s_{\text{LSB}}(t) = A\{[b + \mu m(t)]\cos(2\pi f_c t) + \mu\hat{m}(t)\sin(2\pi f_c t)\}$$

$$s_{\text{USB}}(t) = A\{[b + \mu m(t)]\cos(2\pi f_c t) - \mu\hat{m}(t)\sin(2\pi f_c t)\}$$

where $m(t)$ is the message signal, $\hat{m}(t)$ is the Hilbert transform of $m(t)$, and μ is the modulation index. If $b = 1$, then $s_{\text{LSB}}(t)$ and $s_{\text{USB}}(t)$ become *full-carrier single-sideband* signal. If $b = 0$, the signals become *suppressed-carrier single-sideband* signals. If $0 > b > 1$, the signals become *reduced-carrier single-sideband* signals. The designations *SSB* or *single-sideband* without further qualification are commonly used to indicate suppressed-carrier single-sideband signals.

SSB modulation via frequency discrimination

An obvious "brute force" approach for generating an SSB signal involves creating a DSBSC signal and then removing one of the sidebands by filtering. This approach, sometimes called the *frequency discrimination method* or *filter method*, is most feasible when the baseband message signal is a human voice. Although message signals are often thought of as lowpass signals, the human voice is actually a bandpass signal with a lower cutoff frequency around 70 Hz and an upper cutoff frequency somewhere between 3 and 4 kHz. It has been experimentally determined that the voice spectrum can be further restricted without causing a significant reduction in intelligibility. It is a fairly common practice to restrict speech signals to a band of frequencies between 300 Hz and 3 kHz. As shown in Fig. 12.20, this permits the sideband selection filter to have a transition bandwidth of 600 Hz. If the message signal is truly lowpass, there will be no gap between the two sidebands in the corresponding DSBSC signal. Since any practical realizable filter will have a nonzero transition bandwidth, we are then faced with the choice between either removing some of the desired sideband or leaving some of the undesired sideband. The latter alternative is called *vestigial sideband* (VSB) modulation.

Generation of SSB signals using a Hartley modulator

The *phase discrimination* approach for generation of SSB signals is suggested by the canonical form representations given in Eqs.

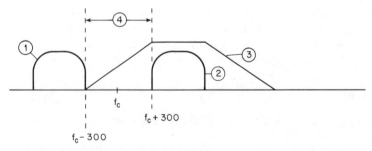

Figure 12.20 Selection of one sideband from a DSBSC-modulated voice signal—rejected sideband (1), selected sideband (2), passband of sideband selection filter (3), and transition band of filter (4).

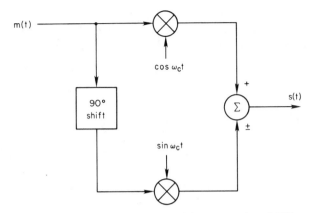

Figure 12.21 Hartley modulator used for generation of SSB signals.

(12.56) and (12.57). This approach employs the *Hartley modulator* shown in Fig. 12.21. Notice the ± symbol where the lower path enters the summer. The lower sideband will be produced when the sine term is added, and the upper sideband will be produced when the sine term is subtracted. As a practical matter, it is difficult to construct the wideband phase shifter shown. A more feasible alternative, shown in Fig. 12.22, includes two phase shifters. The amount of phase shift produced by each shifter need not be constant over frequency so long as a constant 90° is maintained between the two shifter outputs. In theory, this approach is perfect; but in practice the suppression of the undesired sideband is limited by the degree of matching between the

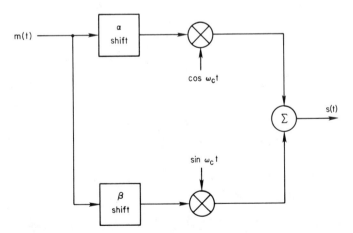

Figure 12.22 Hartley modulator structure which has been modified for easier implementation.

two product modulators as well as by the ability to maintain exact 90° phase shifts between the shifter outputs and between the sine and cosine oscillators. According to Haykin [Ref. 1], 20 dB of suppression is "easy," 30 dB is "reasonable," and more than 40 dB is "quite difficult."

Demodulation of SSB signals [Refs. 4 and 5]

Single-sideband signals can be modulated using the coherent demodulator shown in Fig. 12.15. The successful demodulation of SSB signals depends upon the availability of a coherent reference signal. Usually it will not be possible to make the phase and frequency of the local reference exactly match the carrier. For a frequency error of Δf and a phase error of $\Delta \phi$, the demodulator output will be

$$y(t) = \frac{A}{2}[m(t)\cos(\Delta\omega t + \Delta\phi) \pm \hat{m}(t)\sin(\Delta\omega t + \Delta\phi)]$$

where $\pm$ is taken as $+$ for USB signals and as $-$ for LSB signals.

12.9 Angle Modulation

An angle modulated signal $s(t)$ can be represented as:

$$s(t) = A\,\cos[\theta(t)] \qquad (12.58)$$

where $\theta(t)$ is in some way modulated by the message signal $m(t)$. We can expand $\theta(t)$ as:

$$\theta(t) = \omega_c t + \phi(t) \qquad (12.59)$$

where
$$\omega_c = carrier\ frequency$$
$$\phi(t) = instantaneous\ phase\ deviation$$
$$\omega_c t + \phi(t) = instantaneous\ phase$$
$$\frac{d}{dt}[\omega_c t + \phi(t)] = \omega_c + \frac{d}{dt}\phi(t) = instantaneous\ frequency$$

$$\frac{d}{dt}\phi(t) = instantaneous\ frequency\ deviation$$

In *phase modulation* (PM), the instantaneous phase deviation is made proportional to the modulating signal $m(t)$; thus

$$\phi(t) = K_p m(t) \qquad (12.60)$$

$$s_{\text{PM}}(t) = A\,\cos[\omega_c t + K_p m(t)] \qquad (12.61)$$

The factor K_p is called the *phase sensitivity*. In *frequency modulation* (FM), the instantaneous frequency deviation is made proportional to the modulating signal:

$$\frac{d}{dt}\phi(t) = K_f m(t) \tag{12.62}$$

This is equivalent to making the instantaneous phase deviation proportional to the integral of the modulating signal:

$$\phi(t) = K_f \int m(t)\, dt \tag{12.63}$$

$$s_{\text{FM}}(t) = A \cos\left[\omega_c t + K_f \int m(t)\, dt\right] \tag{12.64}$$

The factor K_f is called the *frequency sensitivity*. Clearly, angle modulation is a nonlinear operation, and consequently the relationship between the spectrum of a message signal and the spectrum of the corresponding FM or PM signal is not nearly as easy to analyze as it is for the case of linear modulation. Traditionally, textbooks presenting the theory of angle modulation have first considered the case where the message signal is a single sinusoid and only then extended these results to cases of multiple sinusoids and general signals. Furthermore, due to the close relationship between FM and PM, results obtained from theoretical analyses of one can be extended to the other by simply integrating or differentiating the message signal.

Single-tone angle modulation

Useful insights into the nature of angle modulation can be gained by examining the case of a sinusoidal modulating signal. Let

$$m(t) = A_m \cos(2\pi f_m t) \tag{12.65}$$

then $$s_{\text{PM}}(t) = A_c \cos[2\pi f_c t + K_p A_m \cos(2\pi f_m t)] \tag{12.66}$$

and $$s_{\text{FM}}(t) = A_c \cos[2\pi f_c t + \beta \sin(2\pi f_m t)] \tag{12.67}$$

where $\beta = K_f A_m / f_m$. Some texts refer to β as the *modulation index*. Using the Bessel function identities given by Eqs. (2.28) and (2.30) we can rewrite (12.66) and (12.67) as:

$$s_{\text{PM}}(t) = A_c \sum_{n=-\infty}^{\infty} J_n(K_p A_m) \cos\left(2\pi f_c t + 2\pi n f_m t + \frac{n\pi}{2}\right) \tag{12.68}$$

$$s_{\text{FM}}(t) = A_c \sum_{n=-\infty}^{\infty} J_n(\beta) \cos(2\pi f_c t + 2\pi n f_m t) \tag{12.69}$$

where $J_n(\cdot)$ denotes the nth-order Bessel function of the first kind. The term involving J_0 is the carrier-frequency component. Inspection of Eqs. (12.68) and (12.69) reveals that $s_{\text{PM}}(t)$ and $s_{\text{FM}}(t)$ will have sideband components at $f = f_c \pm nf_m$ for $n = 1, 2, 3, \ldots$ (Sometimes the sidebands at $f_c \pm nf_m$ are called the *nth-order sidebands*.) Thus we see that an angle-modulated signal has infinite bandwidth even for an extremely narrowband message signal such as a single sinusoid. However, for $K_p A_m$ or β less than unity, the strength of the sidebands decreases rapidly with increasing n.

Double-tone angle modulation

Consider the case of a modulating signal that consists of two sinusoids. The corresponding PM signal will be given by

$$s_{\text{PM}}(t) = A_c \cos[2\pi f_c t + K_1 A_1 \cos(2\pi f_1 t) + K_2 A_2 \cos(2\pi f_2 t)] \quad (12.70)$$

It can be shown [Ref. 9] that this reduces to

$$s_{\text{PM}}(t) = A_c \sum_{n=-\infty}^{\infty} \sum_{m=-\infty}^{\infty} J_n(K_1 A_1) J_m(K_2 A_2)$$

$$\cdot \cos\left[2\pi(f_c + nf_1 + mf_2)t + \frac{(n+m)\pi}{2}\right] \quad (12.71)$$

Multitone angle modulation

The results of single-tone and double-tone angle modulation can be extended to the general case of a PM signal modulated by N sinusoids. Such a signal can be represented as:

$$s_{\text{PM}}(t) = A_c \cos\left[2\pi f_c t + \sum_{n=1}^{N} K_n A_n \cos(2\pi f_n t)\right] \quad (12.72)$$

After some mathematical manipulations, this can be written as

$$s_{\text{PM}}(t) = A_c \sum_{n_1=-\infty}^{\infty} \sum_{n_2=-\infty}^{\infty} \cdots \sum_{n_N=-\infty}^{\infty} \prod_{m=1}^{N} J_{n_m}(K_m A_m)$$

$$\cdot \cos\left[2\pi t\left(f_c + \sum_{m=1}^{N} n_m f_m\right) + \sum_{m=1}^{N} \frac{n_m \pi}{2}\right] \quad (12.73)$$

Narrowband FM

For a sinusoidal modulating signal $m(t) = A_m \cos(2\pi f_m t)$, we can define the following *narrowband* FM signal:

$$s_{\text{NBFM}}(t) = A_c \cos(2\pi f_c t) - \beta A_c \sin(2\pi f_c t) \sin(2\pi f_m t) \quad (12.74)$$

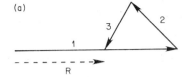

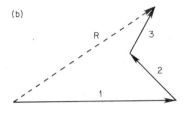

Figure 12.23 Phasor diagrams for an AM signal (a) and the corresponding narrowband FM signal (b).

where $\beta = K_f A_m / f_m$. For values of $\beta \leq 0.3$ rad, Eq. (12.74) is a reasonable approximation to Eq. (12.67). We can expand (12.74) to obtain

$$s_{\text{NBFM}}(t) = A_c \cos(2\pi f_c t) + \frac{1}{2} \beta A_c \cos[2\pi(f_c + f_m)t]$$

$$- \frac{1}{2} \beta A_c \cos[2\pi(f_c - f_m)t] \quad (12.75)$$

The corresponding AM signal can be represented as:

$$s_{\text{AM}}(t) = A_c \cos(2\pi f_c t) + \frac{1}{2} \mu A_c \cos[2\pi(f_c + f_m)t]$$

$$+ \frac{1}{2} \mu A_c \cos[2\pi(f_c - f_m)t] \quad (12.76)$$

The only real difference between Eqs. (12.75) and (12.76) is the sign of the third term in each. The phasor diagram depicted in Fig. 12.23 shows how the sign reversal of just this one component converts the relatively large amplitude variations of the AM signal into phase variations accompanied by very small amplitude variations. The vector components labeled 1, 2, and 3 in the figure represent the first, second, and third terms, respectively, of the corresponding equation.

12.10 References

1. S. Haykin: *Communication Systems*, 2d ed., Wiley, New York, 1983.
2. K. K. Clarke and D. T. Hess: *Communication Circuits: Analysis and Design*, Addison-Wesley, Reading, Mass., 1971.

3. D. C. Green: *Radio Systems for Technicians*, Howard W. Sams, Indianapolis, 1985.
4. H. Taub and D. L. Schilling: *Principles of Communication Systems*, 2d ed., McGraw-Hill, New York, 1986.
5. G. R. Cooper and C. D. McGillem: *Modern Communications and Spread Spectrum*, McGraw-Hill, New York, 1986.
6. R. S. Simpson and R. C. Houts: *Fundamentals of Analog and Digital Communications Systems*, Allyn and Bacon, Boston, 1971.
7. S. Stein and J. S. Jones: *Modern Communication Principles*, McGraw-Hill, New York, 1967.
8. W. Davenport and W. Root: *Random Signals and Noise*, McGraw-Hill, New York, 1958.
9. Technical Staff of Bell Telephone Laboratories: *Transmission Systems for Communications*, Holmdel, N.J., 1964.

Chapter 13

On-Off Keying

The signal set for *on-off keying* (OOK) is often given as:

Space: $s_0(t) = 0 \quad 0 \le t \le T$ (13.1a)

Mark: $s_1(t) = A \cos(2\pi f_c t + \theta) \quad 0 \le t \le T$ (13.1b)

This signal set is an idealized model of OOK in which a mark is a carrier burst that has a rectangular envelope as shown in Fig. 13.1. A more general OOK signal set that admits nonrectangular pulses is given by

$$s_i(t) = \begin{cases} u_T(t) \cos(2\pi f_c t + \theta) & i = 1 \\ 0 & i = 0 \end{cases} \quad (13.2)$$

where $u_T(t)$ is a lowpass waveform that represents some sort of pulse of width T.

13.1 Noncoherent Detection of OOK

Noncoherent detection of OOK signals can be performed using the receiver structure shown in Fig. 13.2. The output of the envelope detector is sampled once per symbol, and bit decisions are made by comparing this sample to a threshold. If the envelope sample r exceeds the threshold λ, it is decided that s_1 has been transmitted. If, on the other hand, $r < \lambda$, it is decided that s_0 has been transmitted. Naturally this decision process is subject to corruption by noise, *intersymbol interference* (ISI), and synchronization errors. Synchronization errors are caused by a mismatch between the symbol rate and the rate at which the output of the envelope detector is sampled.

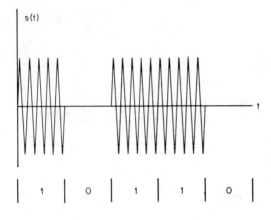

Figure 13.1 Waveform for an OOK signal with rectangular envelope.

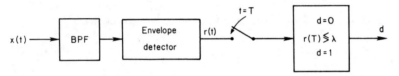

Figure 13.2 Receiver structure for noncoherent detection of OOK signals.

Intersymbol interference occurs whenever the envelope detector is still responding to prior symbol(s) at the onset of a new symbol.

Performance

Assumptions

1. The receiver input $x(t)$ consists of either $s_0(t)$ or $s_1(t)$ as defined above plus stationary additive white gaussian noise.
2. The average noise power at the output of the bandpass filters is $N = \sigma_n^2$.
3. The bandwidths of the channel and of the bandpass filter are sufficiently large such that intersymbol interference does not occur.
4. Adequate measures have been taken to ensure that synchronization errors do not occur.

Analysis. The output of the bandpass filter due to the noise applied to the input is given by

$$n(t) = n_I(t) \cos(2\pi f_c t) - n_Q(t) \sin(2\pi f_c t) \qquad (13.3)$$

Thus when s_0 is being received, the envelope detector output is given by

$$r(t) = \sqrt{n_I^2(t) + n_Q^2(t)}$$

and at any instant a sample of this output will have a Rayleigh pdf (see Sec. 8.3) which is given by

$$p(r \mid s_0) = \frac{r}{\sigma_n^2} \exp\left(\frac{-r^2}{2\sigma_n^2}\right) \quad (13.4)$$

(It should be noted that r is a function of the sampling instant t_0, but in the development that follows, it will be notationally cumbersome to explicitly show this dependence.)

The bandpass filter's output due to $s_1(t)$ applied to the input can be represented as:

$$s(t) = u(t) \cos(2\pi f_c t) \quad (13.5)$$

[*Note*: Since this analysis concerns noncoherent detection, we are free to select an arbitrary phase reference such that θ can be eliminated in (13.5).] The overall output due to signal plus noise is given by

$$x(t) = n(t) + s(t) = [u(t) + n_I(t)] \cos(2\pi f_c t) - n_Q(t) \sin(2\pi f_c t)$$

Thus, when s_1 is being received, the envelope detector output is given by

$$r(t) = \sqrt{[u(t) + n_I(t)]^2 + n_Q^2(t)}$$

and at any instant a sample of this output will have a Rice pdf (see Sec. 8.7) which is given by

$$p(r \mid s_1) = \frac{r}{\sigma_n^2} \exp\left(\frac{r^2 + u^2}{-2\sigma_n^2}\right) I_0\left(\frac{ur}{\sigma_n^2}\right) \quad (13.6)$$

(*Note*: Both r and y are functions of the sampling instant.)

The probability of the receiver erroneously deciding in favor of s_1 when in fact s_0 is being received is equal to the probability that r will exceed λ given that s_0 has been transmitted. This probability is simply that area under that portion of $p(r \mid s_0)$ lying to the right of $r = \lambda$.

$$P_0 \triangleq P(\text{deciding } s_1 \mid s_0 \text{ transmitted})$$

$$= \int_\lambda^\infty p(r \mid s_0)\, dr$$

$$= \exp\left(\frac{-\lambda^2}{2\sigma_n^2}\right) \tag{13.7}$$

The area corresponding to P_0 is shown shaded in Fig. 13.3. [The notation $p(r \mid s_0)$ used in the figure denotes "the probability density of r given that s_0 was transmitted."] The probability of deciding in favor of s_0 when in fact s_1 has been transmitted is equal to the area under that portion of $p(r \mid s_1)$ lying to the left of $r = \lambda$.

$$P_1 \triangleq P(\text{deciding } s_0 \mid s_1 \text{ transmitted})$$

$$= \int_0^\lambda p(r \mid s_1)\, dr = \int_0^\lambda \frac{r}{\sigma_n^2} \exp\left(\frac{r^2 + u^2}{-2\sigma_n^2}\right) I_0\left(\frac{ur}{\sigma_n^2}\right) dr$$

$$= 1 - Q_M\left(\frac{u}{\sigma_n}, \frac{\lambda}{\sigma_n}\right) \tag{13.8}$$

where $Q_M(a, b)$ denotes the Marcum Q function presented in Sec. 2.8. The area corresponding to P_1 is shown shaded in Fig. 13.4. For high

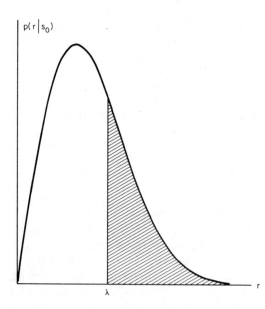

Figure 13.3 Conditional probability density function for the envelope given that a space is transmitted. The shaded area represents the probability of erroneously deciding in favor of a mark.

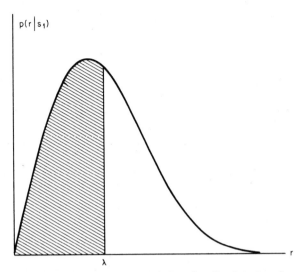

Figure 13.4 Conditional probability density function for the envelope given that a mark is transmitted. The shaded area represents the probability of erroneously deciding in favor of a space.

SNR, P_1 can be approximated as:

$$P_1 = 1 - Q\left(\frac{\lambda - u}{\sigma_n}\right) \qquad (13.9)$$

Assuming that marks and spaces are equally probable, the average probability of error is then given by

$$P_e = \frac{1}{2}P_0 + \frac{1}{2}P_1$$

$$= \frac{1}{2}\exp\left(\frac{-\lambda^2}{2\sigma_n^2}\right) + \frac{1}{2}\left[1 - Q_M\left(\frac{u}{\sigma_n}, \frac{\lambda}{\sigma_n}\right)\right] \qquad (13.10)$$

The probability of error as given in (13.10) is a function of three independent variables—the signal envelope u (at the sampling instant), the decision threshold λ, and the noise variance σ_n^2. Plotting values of P_e is much more convenient if the number of independent variables is reduced to 2 by substituting

$$\gamma = \frac{u^2}{2\sigma_n^2} \qquad \lambda_0 = \frac{\lambda}{\sigma_n} \qquad (13.11)$$

(The quantity γ is just the signal-to-noise ratio at the bandpass filter's output at the sampling instant.) Using the substitutions indicated in

(13.11), the various error probabilities given by (13.7), (13.8), and (13.10) become, respectively,

$$P_0 = \exp\left(\frac{-\lambda_0^2}{2}\right) \tag{13.12}$$

$$P_1 = 1 - Q_M(\sqrt{2\gamma}, \lambda_0) \tag{13.13}$$

$$P_e = \frac{1}{2}\exp\left(\frac{-\lambda_0^2}{2}\right) + \frac{1}{2}[1 - Q_M(\sqrt{2\gamma}, \lambda_0)] \tag{13.14}$$

Plots of P_e versus γ for several values of λ_0 are shown in Fig. 13.5.

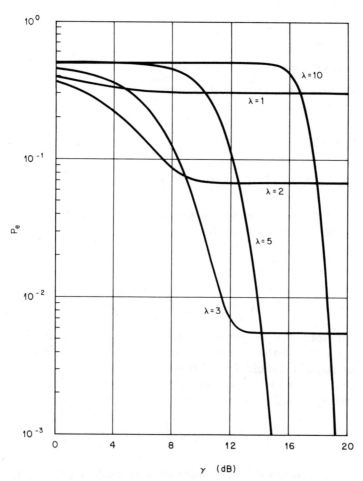

Figure 13.5 Probability of bit error for noncoherent detection of OOK.

Optimum threshold

Examination of Fig. 13.5 reveals that for any given value of λ_0, the value of P_e approaches a constant for high values of the SNR. At these high values of SNR, the error probability P_1 becomes negligible, and the value of P_e is due almost entirely to deciding in favor of s_1 when in fact s_0 is being received. For any given SNR, there will be an optimum value of λ that minimizes the average probability of error when OOK signals are noncoherently decoded. As depicted in Fig. 13.6, this optimum λ will be equal to the value of r at which $p(r \mid s_0) = p(r \mid s_1)$. The optimum threshold can be determined from the following equation:

$$\frac{u^2}{2\sigma_n^2} = \ln I_0\left(\hat{\lambda}\frac{u}{\sigma_n}\right) \quad \text{or} \quad \gamma = \ln I_0(\hat{\lambda}_0\sqrt{2\gamma}) \tag{13.15}$$

A good approximation for the value of $\hat{\lambda}_0$ which solves Eq. (13.15) is given by

$$\hat{\lambda}_0 \approx \left(2 + \frac{u^2}{4\sigma_n^2}\right)^{1/2} = \left(2 + \frac{\gamma}{2}\right)^{1/2} \tag{13.16}$$

A good approximation for the average probability of error attained when the optimum threshold is used is given by

$$P_e \approx \frac{1}{4}\text{erfc}\left(\frac{\sqrt{\gamma}}{2}\right) + \frac{1}{2}\exp\left(\frac{-\gamma}{4}\right) \tag{13.17}$$

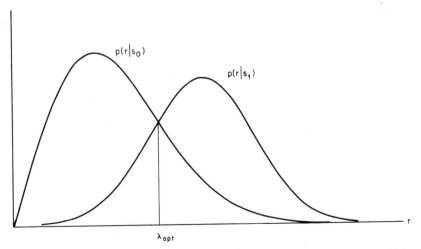

Figure 13.6 The optimum value of λ is equal to the value of r at which $p(r \mid s_0) = p(r \mid s_1)$.

13.2 Coherent Detection of OOK

Coherent detection of OOK signals can be performed using the receiver structure shown in Fig. 13.7. The output of the lowpass filter is sampled once per symbol, and bit decisions are made by comparing this sample to a threshold. If the sample r reveals the threshold λ, it is decided that s_1 has been transmitted. If, on the other hand, $r < \lambda$, it is decided that s_0 has been transmitted.

Performance

Assumptions

1. The receiver input $x(t)$ consists of either $s_0(t)$ or $s_1(t)$ as defined previously, plus stationary additive white gaussian noise.
2. The average noise power at the output of the bandpass filters is $N = \sigma_n^2$.
3. The bandwidths of the channel and of the bandpass filter are sufficiently large such that intersymbol interference does not occur.
4. Adequate measures have been taken to ensure that synchronization errors do not occur.

Analysis. The output of the bandpass filter due to the noise applied to the input is given by

$$n(t) = n_I(t) \cos(2\pi f_c t) - n_Q(t) \sin(2\pi f_c t) \qquad (13.18)$$

The output of the multiplier is then

$$y(t) = n(t) \cos(2\pi f_c t)$$
$$= \frac{1}{2} n_I(t) + \frac{1}{2} n_I(t) \cos(4\pi f_c t) - \frac{1}{2} n_Q(t) \sin(4\pi f_c t) \qquad (13.19)$$

The characteristics of the lowpass filter are selected such that the second and third terms of (13.19) are completely rejected while the

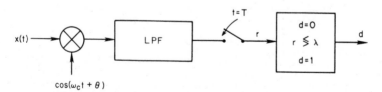

Figure 13.7 Receiver structure for coherent detection of OOK signals.

first term is passed without distortion. Neglecting a constant gain term, the output of the lowpass filter is simply $n_I(t)$. Thus, when s_0 is being received, a sample of the lowpass filter's output at any instant will have a gaussian pdf with zero mean and a variance of σ_n^2.

$$p(r \mid s_0) = \frac{1}{\sqrt{2\pi}\sigma_n} \exp\left(\frac{-r^2}{2\sigma_n^2}\right) \quad (13.20)$$

The bandpass filter's output due to $s_1(t)$ applied at the input can be represented as:

$$s(t) = u(t)\cos(2\pi f_c t) \quad (13.21)$$

and the overall output due to signal plus noise is given by

$$x(t) = n(t) + s(t) = [u(t) + n_I(t)]\cos(2\pi f_c t) - n_Q(t)\sin(2\pi f_c t) \quad (13.22)$$

The output of the multiplier is then

$$y(t) = x(t)\cos(2\pi f_c t)$$
$$= \frac{1}{2}[u(t) + n_I(t)] + \frac{1}{2}[u(t) + n_I(t)]\cos(4\pi f_c t) - \frac{1}{2}n_Q(t)\sin(4\pi f_c t) \quad (13.23)$$

Neglecting a constant gain, the output of the lowpass filter is thus given by

$$r(t) = u(t) + n_I(t)$$

Thus, when s_1 is being received, a sample of the lowpass filter's output at any instant t_0 will have a gaussian pdf with a mean of $u(t_0)$ and a variance of σ_n^2.

$$p(r \mid s_1) = \frac{1}{\sqrt{2\pi}\sigma_n} \exp\left(\frac{-(r-u)^2}{2\sigma_n^2}\right) \quad (13.24)$$

The probability of deciding in favor of s_1 when in fact s_0 has been transmitted is equal to the area under that portion of $p(r \mid s_0)$ lying to the right of $r = \lambda$.

$$P_0 \triangleq P(\text{deciding } s_1 \mid s_0 \text{ transmitted})$$
$$= \int_\lambda^\infty p(r \mid s_0)\, dr$$
$$= \frac{1}{2}\operatorname{erfc}\left(\frac{\lambda}{\sqrt{2}\sigma_n}\right) \quad (13.25)$$

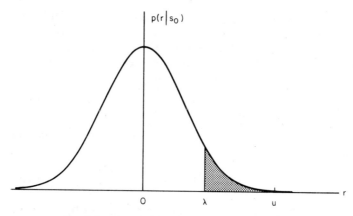

Figure 13.8 Conditional probability density function for the detection variable r given that a space is transmitted. The shaded area represents the probability of erroneously deciding in favor of a mark.

The area corresponding to P_0 is shown in Fig. 13.8. The probability of deciding in favor of s_0 when in fact s_1 has been transmitted is equal to the area under that portion of $p(r \mid s_1)$ lying to the left of $r = \lambda$.

$$P_1 \triangleq P(\text{deciding } s_0 \mid s_1 \text{ transmitted})$$

$$= \int_0^\lambda p(r \mid s_1)\, dr$$

$$= 1 - \frac{1}{2} \operatorname{erfc}\left(\frac{\lambda - u}{\sqrt{2}\sigma_n}\right) \tag{13.26}$$

The area corresponding to P_1 is shown shaded in Fig. 13.9. Assuming

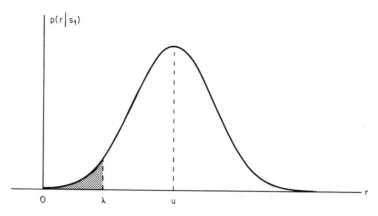

Figure 13.9 Conditional probability density function for the detection variable r given that a mark is transmitted. The shaded area represents the probability of erroneously deciding in favor of a space.

that marks and spaces are equally probable, the average probability of error is given by

$$P_e = \frac{1}{2} + \frac{1}{4}\operatorname{erfc}\left(\frac{\lambda}{\sqrt{2}\sigma_n}\right) - \frac{1}{4}\operatorname{erfc}\left(\frac{\lambda - u}{\sqrt{2}\sigma_n}\right) \qquad (13.27)$$

The probability of error as given in (13.27) is a function of three independent variables—the signal envelope u (at the sampling instant), the decision threshold λ, and the noise variance σ_n^2. Plotting values of P_e is much more convenient if the number of independent variables is reduced to 2 by substituting:

$$\gamma = \frac{u^2}{2\sigma_n^2} \qquad \lambda_0 = \frac{\lambda}{\sigma_n} \qquad (13.28)$$

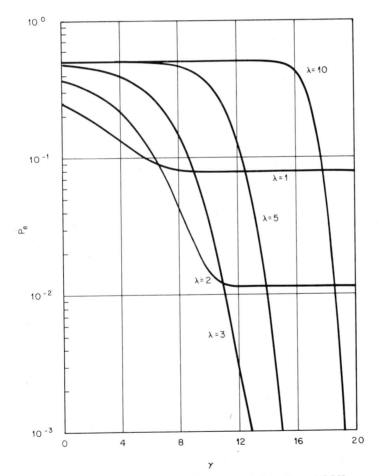

Figure 13.10 Probability of bit error for coherent detection of OOK.

(The quantity γ is just the signal-to-noise ratio at the bandpass filter's output at the sampling instant.) Using the substitutions indicated in (13.28), the various error probabilities given by (13.25), (13.26), and (13.27) become, respectively,

$$P_0 = \frac{1}{2}\operatorname{erfc}\left(\frac{\lambda_0}{\sqrt{2}}\right) \tag{13.29}$$

$$P_1 = 1 - \frac{1}{2}\operatorname{erfc}\left(\frac{\lambda_0}{\sqrt{2}} - \gamma\right) \tag{13.30}$$

$$P_e = \frac{1}{2} + \frac{1}{4}\operatorname{erfc}\left(\frac{\lambda_0}{\sqrt{2}}\right) - \frac{1}{4}\operatorname{erfc}\left(\frac{\lambda_0}{\sqrt{2}} - \sqrt{\gamma}\right) \tag{13.31}$$

Plots of P_e versus γ for several values of λ_0 are shown in Fig. 13.10.

Optimum threshold

There is an optimum value of λ that minimizes the average probability of error. Unlike the noncoherent case, this threshold depends only upon the signal strength, rather than upon the SNR. The optimum threshold will be equal to the value of r at which $p(r \mid s_0) = p(r \mid s_1)$. Since the variance of the two distributions is equal, the optimum threshold is simply halfway between the two means; thus

$$\hat{\lambda} = \frac{u(t_0)}{2} \quad \text{or} \quad \hat{\lambda}_0 = \frac{u(t_0)}{2\sigma_n} = \sqrt{\frac{\gamma}{2}} \tag{13.32}$$

When coherent detection is performed using the optimum threshold, the probability of error is given by

$$P_e = \frac{1}{2}\operatorname{erfc}\left(\frac{u}{2^{3/2}\sigma_n}\right) = \frac{1}{2}\operatorname{erfc}\left(\frac{\sqrt{\gamma}}{2}\right) \tag{13.33}$$

13.3 References

1. S. Stein and J. S. Jones: *Modern Communications Principles*, McGraw-Hill, New York, 1967.
2. M. Schwartz, W. R. Bennett, and S. Stein: *Communication Systems and Techniques*, McGraw-Hill, New York, 1966.
3. G. R. Cooper and C. D. McGillem: *Modern Communications and Spread Spectrum*, McGraw-Hill, New York, 1986.

Chapter 14

Frequency-Shift Keying

14.1 Binary Frequency-Shift Keying

A signal set for binary *frequency-shift keying* (FSK) can be defined as:

$$s_0(t) = A \cos\left[2\pi\left(f_c + \frac{\Delta f}{2}\right)t + \theta_0\right] \quad 0 < t < T \quad (14.1a)$$

$$s_1(t) = A \cos\left[2\pi\left(f_c - \frac{\Delta f}{2}\right)t + \theta_1\right] \quad 0 < t < T \quad (14.1b)$$

In order to explicitly show the carrier frequency f_c, Eq. (14.1a) can be rewritten as:

$$s_0(t) = A \cos\left[2\pi\left(\frac{f_c + \Delta f}{2}\right)t + \theta_0\right] \quad 0 < t < T \quad (14.2a)$$

$$s_1(t) = A \cos\left[2\pi\left(\frac{f_c - \Delta f}{2}\right)t + \theta_1\right] \quad 0 < t < T \quad (14.2b)$$

where $\Delta f = f_0 - f_1 \quad f_c = \frac{f_0 + f_1}{2}$

The complete transmitted signal $s(t)$ can be expressed as:

$$s(t) = A \sum_{n=0}^{\infty} \cos\left[2\pi\left(f_c + \frac{b_n \Delta f}{2}\right) + \phi_n\right] g(t - nT) \quad (14.3)$$

where $\{b_n\}$ denotes the random sequence of bits to be transmitted, with each individual b_n value selected from the set $\{-1, 1\}$, and ϕ_n given by

$$\phi_n = \begin{cases} \theta_1 & b_n = 1 \\ \theta_0 & b_n = -1 \end{cases}$$

The equivalent baseband pulse shape for bit n is denoted as $g(t - nT)$, which for rectangular pulses is given by

$$g(t - nT) = \begin{cases} 1 & nT < t \leq (n+1)T \\ 0 & \text{otherwise} \end{cases}$$

Equation (14.3) can also be expressed in terms of the complex envelope $\tilde{s}(t)$:

$$s(t) = Re[\tilde{s}(t) \exp(j2\pi f_c t)] \qquad (14.4)$$

where $\tilde{s}(t) = A \sum_{n=0}^{\infty} \exp(j\pi \Delta f t b_n + \phi_n) g(t - nT) \qquad (14.5)$

The FSK signal defined by (14.3) has a spectral density given by

$$\begin{aligned} S(f) = & \frac{A}{\theta} \delta(f - f_0) + \frac{A}{\theta} \delta(f - f_1) + \frac{N(f - f_0)}{D^2(f - f_0)} \\ & + \frac{N(f + f_0)}{D^2(f + f_0)} + \frac{N(f - f_1)}{D^2(f - f_1)} + \frac{N(f + f_1)}{D^2(f + f_1)} \\ & + \mu(2f_0) \frac{2\cos(2\theta_0) N(f - f_0)}{4\pi^2 (f^2 - f_0^2)} + \mu(2f_1) \frac{2\cos(2\theta_1) N(f - f_1)}{4\pi^2 (f^2 - f_1^2)} \\ & - 2\mu(f_1 + f_0) \cos(\theta_1 + \theta_0) \left[\frac{N(f - f_0)}{D(f - f_0) D(f + f_1)} + \frac{N(f - f_1)}{D(f + f_0) D(f - f_1)} \right] \\ & - 2\mu(f_1 - f_0) \cos(\theta_1 - \theta_0) \left[\frac{N(f - f_0)}{D(f - f_0) D(f - f_1)} + \frac{N(f + f_1)}{D(f + f_0) D(f + f_1)} \right] \end{aligned}$$

$$(14.6)$$

where $N(f) = \sin^2(\pi f t)$
$D(f) = 2\pi f$
$$\mu(f) = \begin{cases} 1, & fT = \text{an integer} \\ 0, & fT \neq \text{an integer} \end{cases}$$

The spectrum defined by (14.6) is plotted in Fig. 14.1 for several values of $|f_0 - f_1|$. For large values of f_0 and f_1 with $(f_1 - f_0)$ not an integer multiple of $1/T$, Eq. (14.6) can be approximated as:

$$S(t) = \frac{A}{8} \delta(f - f_0) + \frac{A}{8} \delta(f - f_1) + \frac{A^2 N(f - f_0)}{2TD^2(f - f_0)} + \frac{A^2 N(f - f_1)}{2TD^2(f - f_1)} \qquad (14.7)$$

The spectral density for FSK signals is derived in Refs. 4 and 5.

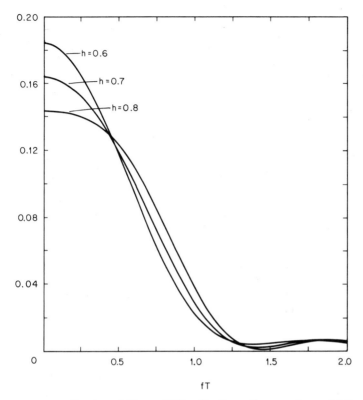

Figure 14.1 Spectra of binary FSK with discontinuous phase at bit transitions.

14.2 Continuous-Phase Frequency-Shift Keying (CPFSK)

The FSK signals presented in Sec. 14.1 exhibit phase discontinuities at the transitions between bits as shown in Fig. 14.2, and these discontinuities give rise to relatively large spectral sidelobes. These undesired sidelobes can be avoided by using an alternative form of FSK called *continuous-phase* FSK (CPFSK) which switches smoothly from bit to bit as shown in Fig. 14.3. A complete transmitted signal for CPFSK can be defined as:

$$s(t) = Re[\tilde{s}(t) \exp(j2\pi f_c t)] \quad (14.8)$$

The complex envelope $\tilde{s}(t)$ is given by

$$\tilde{s}(t) = A \sum_{n=0}^{\infty} \exp\left\{j\phi + j2\pi f_d \left[\left(\sum_{k=0}^{n-1} b_k\right)T + (t - nT)b_n\right]\right\} g(t - nT) \quad (14.9)$$

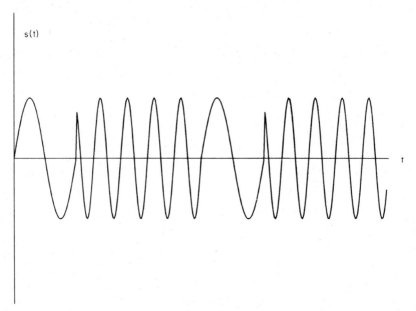

Figure 14.2 Typical waveform for binary FSK with discontinuous phase at bit transitions.

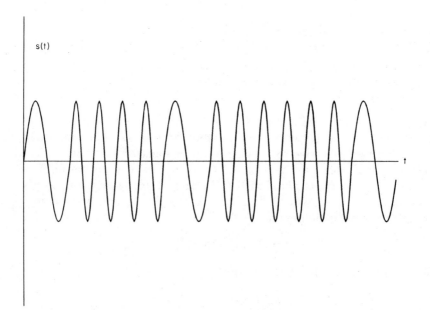

Figure 14.3 Typical waveform for binary CPFSK.

where ϕ = arbitrary (usually assumed random or zero) initial phase
f_d = peak frequency deviation
b_n = sequence of bits to be transmitted
$b_n \in \{1, -1\}$

$$g(t - nT) = \begin{cases} 1 & nT < t \le (n+1)T \\ 0 & \text{otherwise} \end{cases}$$

It is not particularly useful to define a signal set which represents CPFSK over just a single bit interval since the phase at the beginning of the bit interval must equal the phase at the end of the previous bit interval, which equals the sum of the phase at time zero plus the total accumulated phase since time zero.

The binary CPFSK signal $s(t)$ defined by Eq. (14.8) has a spectral density given by one of two different expressions depending upon whether the magnitude of the characteristic function $\psi(j2\pi f_d T)$ is less than or equal to unity. For binary CPFSK, the characteristic function is given by

$$\psi(j2\pi f_d T) = \cos(2\pi f_d T) \tag{14.10}$$

Thus $|\psi| = 1$ for all values of f_d that are integer multiples of $1/(2T)$. For cases where $|\psi| < 1$, it can be shown [Ref. 6] that the spectrum of $s(t)$ is given by

$$S(f) = \frac{T}{8}\{A_1^2(f)[1 + B_2(f)] + A_2^2(f)[1 + B_4(f)] + 2A_1(f)A_2(f)B_3(f)\} \tag{14.11}$$

where $A_1(f) = \dfrac{\sin[\pi T(f + f_d)]}{\pi T(f + f_d)}$

$A_2(f) = \dfrac{\sin[\pi T(f - f_d)]}{\pi T(f - f_d)}$

$B_n(f) = \dfrac{\cos(2\pi fT - \alpha_n) - \beta \cos \alpha_n}{1 + \beta^2 - 2\beta \cos(2\pi fT)}$

$\alpha_n = 2\pi f_d T(n - 3)$

$\beta = \psi(j2\pi f_d T)$

The spectrum defined by (14.11) is plotted in Figs. 14.4 thru 14.7 for various values of f_d.

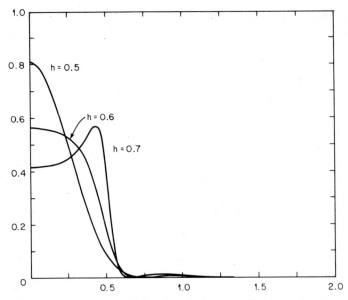

Figure 14.4 Power spectral density of binary CPFSK signals with $h = 0.5$, $h = 0.6$, and $h = 0.7$.

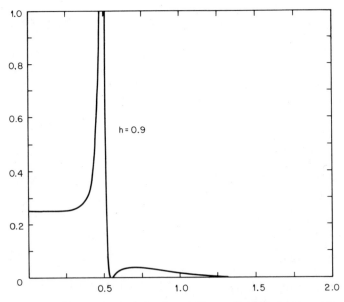

Figure 14.5 Power spectral density of binary CPFSK signals with $h = 0.9$.

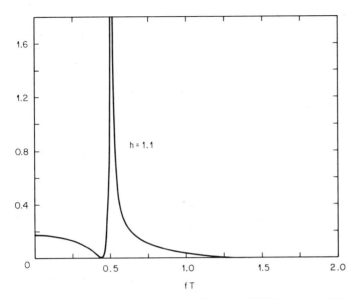

Figure 14.6 Power spectral density of binary CPFSK signals with $h = 1.1$.

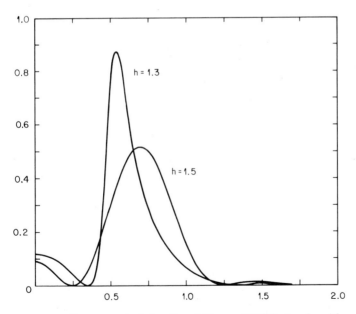

Figure 14.7 Power spectral density of binary CPFSK signals with $h = 1.3$ and $h = 1.5$.

14.3 Coherent Detection of FSK Signals

The receiver structure shown in Fig. 14.8 can be used to perform coherent detection of FSK signals. This structure is simply the structure given in Fig. 10.6, tailored for the FSK signal set. The difference between the two integrator outputs is sampled at the end of each bit interval, and bit decisions are made based upon the sign of this sample. If the difference is positive, it is decided that s_0 has been received. If, on the other hand, the difference is negative, it is decided that s_1 has been received. The receiver input $x(t)$ consists of either $s_0(t)$ or $s_1(t)$ as defined by Eq. (14.1a) plus stationary additive white gaussian noise having a two-sided spectral density equal to $N_0/2$. The bit error analysis presented below assumes that each of the local reference signals provided to the multipliers is exactly phase- and frequency-matched to one of the two signals $s_0(t)$ and $s_1(t)$. It is further assumed that the bandwidths of the channel and of the receiver circuits are sufficiently large such that intersymbol interference does not occur.

Orthogonal signals

Assume that the signals s_0 and s_1 are orthogonal. Therefore, in the absence of noise, $r_1 = 0$ when s_0 is being received, and $r_0 = 0$ when s_1 is being received. The signals s_0 and s_1 will be exactly orthogonal when $|f_1 - f_0| = n/(2T)$, and at least approximately orthogonal when $|f_1 - f_0| \gg 1/T$. For s_0 and s_1 orthogonal, it follows directly from Eq. (10.28) that the probability of error is given by

$$P_e = \frac{1}{2}\operatorname{erfc}\left[\left(\frac{E_b}{2N_0}\right)^{1/2}\right] \qquad (14.12)$$

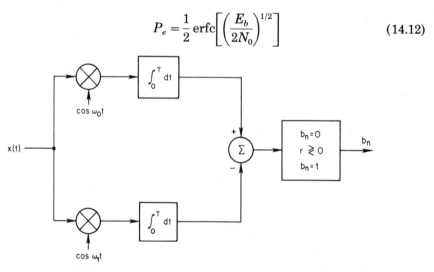

Figure 14.8 Receiver structure for coherent detection of binary FSK signals.

where $E_b = A^2T/2$. For high SNR, the error probability can be approximated as:

$$P_e \approx \left(\frac{N_0}{2\pi E_b}\right)^{1/2} \exp\left(\frac{E_b}{2N_0}\right) \qquad (14.13)$$

Nonorthogonal signals

Assume that the signals are not orthogonal but that the FSK tone frequencies are large enough relative to the bit rate such that

$$f_1 + f_0 \gg \frac{1}{T} \qquad (14.14)$$

Under these conditions, the probability of error is given by

$$P_e = \frac{1}{2}\text{erfc}\left\{\left[(1-\rho)\frac{E_b}{2N_0}\right]^{1/2}\right\} \qquad (14.15)$$

where $\rho \approx \dfrac{\sin[2\pi(f_1 - f_0)T]}{2\pi(f_1 - f_0)T}$ \qquad (14.16)

If (14.14) is not satisfied, the approximation in (14.16) is not valid and ρ must be computed as:

$$\rho = \frac{\sin[2\pi(f_1 - f_0)T]}{2\pi(f_1 - f_0)T} - \frac{\sin[2\pi(f_1 + f_0)T]}{2\pi(f_1 + f_0)T} \qquad (14.17)$$

The correlation ρ will be a minimum for f_1 and f_0 that satisfy

$$|f_1 - f_0| = \frac{0.71}{T} \qquad (14.18)$$

For signals that satisfy (14.18), the probability of error will be at a minimum given by

$$P_e = \frac{1}{2}\text{erfc}\left[\left(0.6086\,\frac{E_b}{N_0}\right)^{1/2}\right] \qquad (14.19)$$

Note also that for $|f_1 - f_0| = n/(2T)$, Eq. (14.15) reduces to Eq. (14.12). The probability of error defined by (14.15) is plotted against E_b/N_0 for various values of ρ in Fig. 14.9.

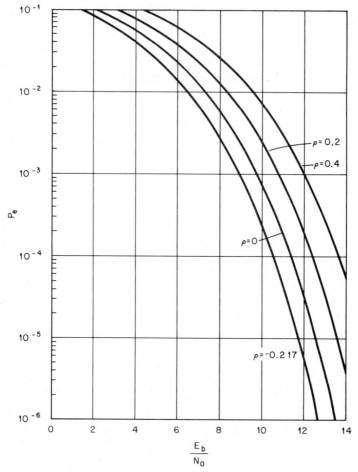

Figure 14.9 Probability of bit error versus E_b/N_0 for coherent detection of binary FSK signals.

14.4 Noncoherent Detection of FSK

Bandpass filter approach

The receiver structure shown in Fig. 14.10 can be used to perform noncoherent detection of FSK signals. The receiver input $x(t)$ will consist of either $s_0(t)$ or $s_1(t)$ as defined by Eq. (14.1a), plus stationary additive white gaussian noise. The power spectral density of the noise is such that the noise power out of each bandpass filter is $N = \sigma_n^2$. Assume that the signals s_0 and s_1 are orthogonal. Therefore, in the absence of noise, $r_1 = 0$ when s_0 is being received, and $r_0 = 0$ when s_1 is being received. The signal s_0 and s_1 will be exactly orthogonal

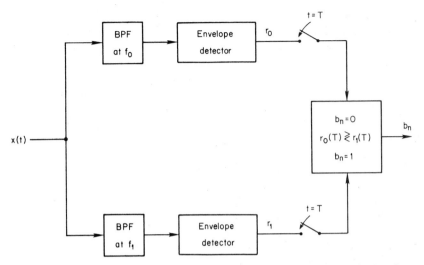

Figure 14.10 Receiver structure for noncoherent detection of binary FSK signals.

when $|f_1 - f_0| = n/(2T)$, and at least approximately orthogonal when $|f_1 - f_0| \gg 1/T$. It is further assumed that the bandwidths of the channel and of the receiver circuits are sufficiently large such that intersymbol interference does not occur.

It follows from the results of Sec. 8.7 that when the signal $s_i(t)$ is being received, the corresponding envelope detector output $r_i(t)$ will have a Rice pdf given by

$$p(r_i \mid s_i) = \frac{r_i}{N} I_0\left(\frac{Ar_i}{N}\right) \exp\left(\frac{r_i^2 + A^2}{-2N}\right) \qquad (14.20)$$

The envelope detector corresponding to the signal which is not being sent will have a Rayleigh pdf given by

$$p(r_j \mid s_i) = \frac{r_j}{N} \exp\left(\frac{-r_j^2}{2N}\right) \qquad (14.21)$$

The probability of erroneously deciding in favor of $s_j(t)$ when in fact $s_i(t)$ is being received is equal to the probability that r_j will exceed r_i given that $s_i(t)$ is being received.

$$P_e = P(r_j > r_i \mid s_i) = \int_{r_i=0}^{\infty} p(r_i \mid s_i) \left[\int_{r_j=r_i}^{\infty} p(r_j \mid s_i) \, dr_j\right] dr_i \qquad (14.22)$$

The inner integral of (14.22) is simply the probability that r_j exceeds any given value of r_i. This probability is a function of r_i and thus must be averaged over all values of r_i to obtain the probability of error P_e. After appropriate mathematical manipulations, Eq. (14.22) reduces to

$$P_e = \frac{1}{2} \exp\left(\frac{-A^2}{4N}\right) \qquad (14.23)$$

Matched-filter approach

The receiver structure shown in Fig. 14.11 will be used to perform noncoherent detection of FSK signals. The characteristics of the signals and noise are the same as given above for the bandpass filter approach. Under these conditions, it can be shown [Refs. 1 through 3] that the probability of error is given by

$$P_e = \frac{1}{2} \exp\left(\frac{-A^2 T}{4N_0}\right) = \frac{1}{2} \exp\left(\frac{-E_b}{2N_0}\right) \qquad (14.24)$$

where $E_b = A^2 T/2$. If we assume that the bandpass filters in Fig. 14.10 each have a noise bandwidth of $1/T$, then Eqs. (14.23) and (14.24) yield identical results.

Nonorthogonal signals

The receiver structure shown in Fig. 14.11 will be used to perform noncoherent detection of FSK signals. The receiver input $x(t)$ will consist of either $s_0(t)$ or $s_1(t)$ as defined by Eq. (14.1a), plus stationary

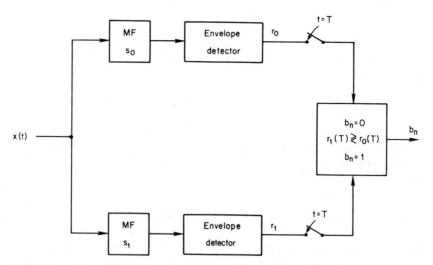

Figure 14.11 Matched-filter receiver structure for noncoherent detection of binary FSK signals.

additive white gaussian noise having a two-sided spectral density equal to $N_0/2$. The signals $s_0(t)$ and $s_1(t)$ are not assumed to be orthogonal. It follows from Eq. (10.41) that the probability of error is given by

$$P_e = Q(a, b) - \frac{1}{2} \exp\left(\frac{a^2 + b^2}{-2}\right) I_0(ab) \qquad (14.25)$$

where $a = \left\{ \frac{E_b}{2N_0} [1 - (1 - |\rho|^2)^{1/2}] \right\}^{1/2}$

$b = \left\{ \frac{E_b}{2N_0} [1 + (1 - |\rho|^2)^{1/2}] \right\}^{1/2}$

$Q(\cdot, \cdot)$ = Marcum Q function (see Sec. 2.8)
$I_0(\cdot)$ = modified Bessel function of order zero (see Sec. 2.7)
ρ = cross-correlation coefficient for s_0 and s_1 which is given by

$$\rho = \frac{\sin(\pi T \, \Delta f)}{\pi T \, \Delta f} \exp(-j\pi T \, \Delta f)$$

$$\Delta f = f_0 - f_1$$

Note that when $\rho = 0$, Eq. (14.25) reduces to (14.24).

14.5 M-ary FSK

The concept of M-ary FSK (MFSK) is a straightforward extension of binary FSK. A signal set for MFSK can be defined as:

$$s_1(t) = A \cos(2\pi f_1 t + \theta_1) \qquad 0 < t < T$$
$$s_2(t) = A \cos(2\pi f_2 t + \theta_2) \qquad 0 < t < T$$
$$\vdots$$
$$s_M(t) = A \cos(2\pi f_M t + \theta_M) \qquad 0 < t < T \qquad (14.26)$$

In general, the frequencies $f_1, f_2, \ldots, f_M$ may be arbitrarily selected, but a case of major practical interest occurs when M is an integer power of 2 and the frequencies are equally spaced:

$$M = 2^N$$
$$f_2 = f_1 + \Delta f$$
$$f_3 = f_2 + \Delta f = f_1 + 2\Delta f$$
$$\vdots$$
$$f_M = f_{M-1} + \Delta f = f_1 + (M - 1) \, \Delta f \qquad (14.27)$$

For this special case, Eq. (14.26) can be rewritten so as to explicitly show the carrier frequency f_c

$$s_i(t) = A \cos\left\{2\pi\left[f_c + \frac{\Delta f(2i - 1 - M)}{2}\right]t + \theta_i\right\} \quad (14.28)$$

where $f_c = \dfrac{f_1 + f_M}{2}$

The complete transmitted signal $s(t)$ can be expressed as:

$$s(t) = A \sum_{n=0}^{\infty} \cos\left[2\pi\left(f_c + \frac{d_n \Delta f}{2}\right) + \phi_n\right] g(t - nT) \quad (14.29)$$

where $\{d_n\}$ denotes the random sequence of symbols to be transmitted, with each individual d_n value selected from the set $\{-M+1, -M+3, \ldots, -1, 1, \ldots, M-3, M-1\}$, and ϕ_n given by

$$\phi_n = \begin{cases} \theta_1 & d_n = -M+1 \\ \theta_2 & d_n = -M+3 \\ \vdots & \\ \theta_{M/2} & d_n = -1 \\ \theta_{1+M/2} & d_n = 1 \\ \vdots & \\ \theta_M & d_n = M-1 \end{cases}$$

The MFSK signal defined by (14.29) will generally exhibit phase discontinuities at the transitions between adjacent symbols. Just as for binary FSK, we can redefine the M-ary FSK signal such that the phase is continuous. A complete transmitted signal for M-ary CPFSK can then be represented as:

$$s(t) = Re[u(t) \exp(j2\pi f_c t)] \quad (14.30)$$

The complex envelope $u(t)$ is given by

$$u(t) = A \sum_{n=0}^{\infty} \exp\left\{j\phi + j2\pi f_d\left[\left(\sum_{k=0}^{n-1} d_k\right)T + (t - nT)d_n\right]\right\} g(t - nT) \quad (14.31)$$

where ϕ = arbitrary (usually assumed random) initial phase
f_d = peak frequency deviation
d_n = as defined for (14.29)

The M-ary CPFSK signal defined by Eq. (14.30) has a spectral density given by one of two different expressions depending upon whether the magnitude of the characteristic function $\psi(j2\pi f_d T)$ is less than or equal to unity. For M-ary CPFSK, the characteristic function is given by

$$\psi(j2\pi f_d T) = \frac{2}{M} \sum_{n=1}^{M/2} \cos[2\pi f_d (2n-1)T] \qquad (14.32)$$

For cases where $|\psi| < 1$, it can be shown [Ref. 6] that the spectrum of $s(t)$ is given by

$$S(f) = \frac{T}{4M} \left[\sum_{n=1}^{M} A_n^2(f) + \frac{2}{M} \sum_{n=1}^{M} \sum_{m=1}^{M} A_n(f) A_m(f) B_{n+m}(f) \right] \qquad (14.33)$$

where $A_n(f) = \dfrac{\sin\{\pi T[f - f_d(2n-1-M)]\}}{\pi T[f - f_d(2n-1-M)]}$

$B_n(f) = \dfrac{\cos(2\pi f T - \alpha_n) - \beta \cos \alpha_n}{1 + \beta^2 - 2\beta \cos(2\pi f T)}$

$\alpha_n = 2\pi f_d T(n - 1 - M)$

$\beta = \psi(j2\pi f_d T)$

The spectrum defined by (14.33) is plotted in Figs. 14.12 through

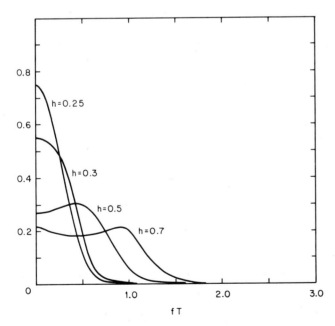

Figure 14.12 Power spectral density for 4-ary CPFSK, with $h = 0.25$, $h = 0.3$, $h = 0.5$, and $h = 0.7$.

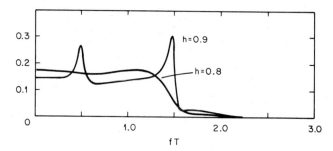

Figure 14.13 Power spectral density for 4-ary CPFSK, with $h = 0.8$ and $h = 0.9$.

14.17 for $M = 4$ and $M = 8$. These spectra are plotted as functions of the normalized frequency fT for various values of the *frequency deviation ratio* $h = 2f_d T$.

14.6 Detection of MFSK Signals

Coherent detection

The receiver structure shown in Fig. 14.18 will be used to perform coherent detection of $MFSK$ signals. The receiver input consists of one of the M possible $s_i(t)$ as defined by Eq. (14.26), plus stationary additive white gaussian noise having a two-sided spectral density equal to $N_0/2$. Assume that the signals $s_i(t)$ are orthogonal and that adequate measures have been taken to ensure that synchronization errors and intersymbol interference do not occur. It then follows

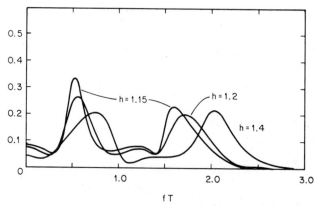

Figure 14.14 Power spectral density for 4-ary CPFSK, with $h = 1.15$, $h = 1.2$, and $h = 1.4$.

Frequency-Shift Keying 235

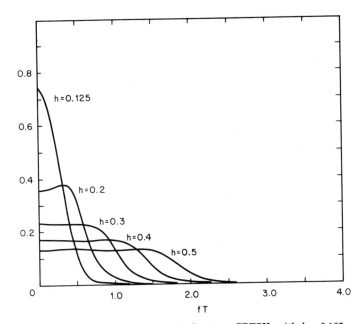

Figure 14.15 Power spectral density for 8-ary CPFSK, with $h = 0.125$, $h = 0.2$, $h = 0.3$, $h = 0.4$, and $h = 0.5$.

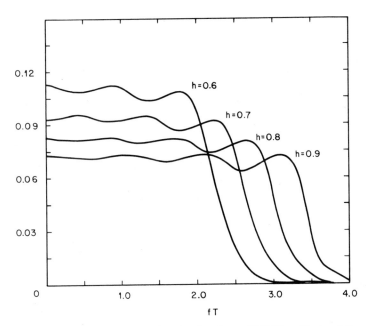

Figure 14.16 Power spectral density for 8-ary CPFSK, with $h = 0.6$, $h = 0.7$, $h = 0.8$, and $h = 0.9$.

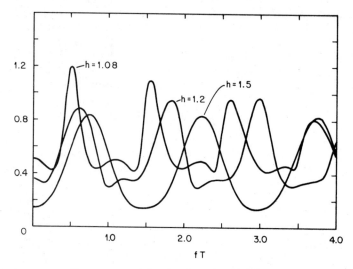

Figure 14.17 Power spectral density for 8-ary CPFSK, with $h = 1.08$, $h = 1.2$, and $h = 1.5$.

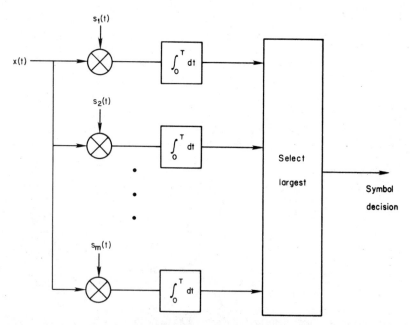

Figure 14.18 Receiver structure for coherent detection of $MFSK$ signals.

directly from Eq. (10.34) that the probability of symbol error is given by

$$P_s = \frac{1}{\sqrt{2\pi}} \int_{-\infty}^{\infty} \left\{ 1 - \left[1 - \frac{1}{2} \text{erfc}\left(\frac{x}{\sqrt{2}}\right) \right]^{M-1} \right\} \cdot \exp\left[\frac{-\left(x - \sqrt{\frac{2kE_b}{N_0}}\right)^2}{2} \right] dx \quad (14.34)$$

where $E_b = \dfrac{A^2 T}{2k}$

k = number of bits per symbol = $\log_2 M$

The probability of bit error can be obtained from P_s using

$$P_b = \frac{2^{k-1}}{2^k - 1} P_s \quad (14.35)$$

Noncoherent detection

The receiver structure shown in Fig. 14.19 will be used to perform noncoherent detection of $MFSK$ signals. The receiver input consists of one of the M possible $s_i(t)$ as defined by Eq. (14.3) plus stationary additive white gaussian noise having a two-sided spectral density equal to $N_0/2$. The signals $s_i(t)$ are orthogonal. Assume that adequate measures have been taken to ensure that synchronization errors and intersymbol interference do not occur.

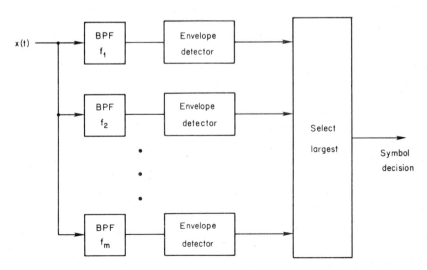

Figure 14.19 Receiver structure for noncoherent detection of $MFSK$ signals.

It follows directly from Eq. (10.42) that the probability of symbol error is given by

$$P_s = \sum_{n=1}^{M-1} (-1)^{n+1} \frac{(M-1)!}{(n+1)!(M-n-1)!} \exp\left(\frac{-nkE_b}{(n+1)N_0}\right) \quad (14.36)$$

where $E_b = \dfrac{A^2 T}{2k}$

k = number of bits per symbol = $\log_2 M$

The probability of bit error can be obtained from P_s using Eq. (14.35). The probability of bit error versus E_b/N_0 is plotted in Fig. 14.20 for several values of M.

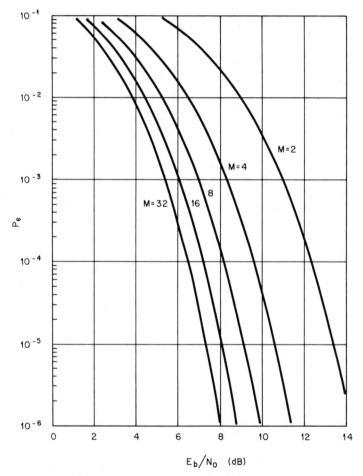

Figure 14.20 Probability of bit error for noncoherent detection of orthogonal MFSK.

14.7 Binary FSK Performance in Rayleigh Fading

Assume that for each bit, the transmitted signal $s(t)$ consists of either $s_0(t)$ or $s_1(t)$ as defined by Eq. (14.1a), plus additive white gaussian noise having a two-sided spectral density equal to $N_0/2$. Further assume that the frequencies f_1 and f_0 are selected such that the signals s_0 and s_1 are orthogonal. Between the transmitter and receiver, the signal $s(t)$ is subjected to Rayleigh fading. The representation of an FSK signal as given by Eq. (14.4) can be changed to incorporate the effects of Rayleigh fading as follows:

$$r(t) = Re[\tilde{r}(t) \exp(j2\pi f_c t)] \tag{14.37}$$

where $\tilde{r}(t) = \alpha \exp(-j\theta) A \sum_{k=0}^{\infty} \exp(j\pi \Delta f t b_k) g(t - kT)$

with α being Rayleigh distributed and θ being uniformly distributed on $[0, 2\pi)$. It is further assumed that the bandwidths of the channel and of the receiver circuits are sufficiently large such that intersymbol interference does not occur.

Coherent detection

The received FSK signal defined by (14.37) will be subjected to coherent detection using the receiver structure shown in Fig. 14.8. It is assumed that the phase estimation needed to generate the local reference can be performed without error. It can be shown [Ref. 6] that the probability of bit error is then given by

$$P_e = \frac{1}{2}\left(1 - \sqrt{\frac{\gamma}{1+\gamma}}\right) \tag{14.38}$$

where $\gamma = \dfrac{E_b}{N_0} E[\alpha^2]$

In other words, γ is just the average ratio of signal energy per bit to noise density, *after* fading of the signal is taken into account.

Noncoherent detection

The received FSK signal defined by (14.37) will be subjected to noncoherent detection using the receiver structure shown in Fig. 14.11. It can be shown that the probability of bit error is given by

$$P_e = \frac{1}{2 + (E_b/N_0)E[\alpha^2]} \tag{14.39}$$

14.8 Diversity Performance of Binary Orthogonal FSK

Assume that for each bit, the transmitted signal $s(t)$ consists of either $s_0(t)$ or $s_1(t)$ as defined by Eq. (14.1a), plus additive white gaussian noise having a two-sided spectral density equal to $N_0/2$. Further assume that the signals s_0 and s_1 are orthogonal. The signal $s(t)$ is transmitted over N different, frequency nonselective, slowly fading channels whose fading processes are assumed to be mutually statistically independent. The signal received via the nth channel can be represented as:

$$r_n(t) = Re[\tilde{r}_n(t) \exp(j2\pi f_c t)] \qquad (14.40)$$

where $\tilde{r}_n(t) = \alpha_n \exp(-j\theta_n) A \sum_{k=0}^{\infty} \exp(j\pi \Delta f t b_k) g(t - kT)$ (14.41)

with α_n being Rayleigh distributed and θ_n being uniformly distributed on $[0, 2\pi)$. The number of independent channels N is called the *diversity order*. The received FSK signal defined by (14.40) will be subjected to noncoherent detection and diversity combining using the receiver structure shown in Fig. 14.21. The probability of bit error is given by

$$P_e = \left(\frac{1-\beta}{2}\right)^N \sum_{n=0}^{N} \frac{(N-1+n)!}{n!(N-1)!} \left(\frac{1+\beta}{2}\right)^n \qquad (14.42)$$

where $\beta = \dfrac{\gamma}{2+\gamma}$

$$\gamma = \frac{E_b}{N_0} E[\alpha^2] = \frac{E_b}{N_0} E[\alpha_n^2] \qquad n = 1, 2, \ldots, N$$

14.9 M-ary FSK Performance in Rayleigh Fading

The receiver structure shown in Fig. 14.19 will be used to perform noncoherent detection of M-ary FSK signals. For each symbol, the transmitted signal $s(t)$ consists of one of the M possible $s_i(t)$ as defined by Eq. (14.28), plus additive white gaussian noise having a two-sided spectral density equal to $N_0/2$. Between the transmitter and the receiver, the signal $s(t)$ will be subjected to Rayleigh fading.

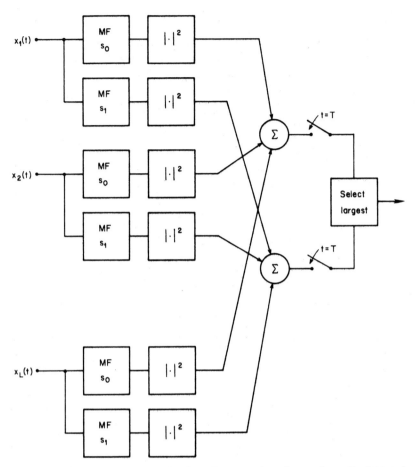

Figure 14.21 Receiver structure for diversity reception of noncoherently detected FSK signals.

Under the stated conditions, the probability of symbol error is given by

$$P_s = \sum_{m=1}^{M-1} \frac{(-1)^{m+1}(M-1)!}{m!(M-m-1)!(1+m+m\gamma)} \tag{14.43}$$

where $\gamma = \dfrac{E_b}{N_0} E[\alpha^2] \log_2 M$

The probability of bit error can be obtained from P_s using Eq. (14.35).

14.10 References

1. R. Gagliardi: *Introduction to Communications Engineering*, Wiley, New York, 1978.
2. S. Haykin: *Communication Systems*, 2d ed., Wiley, New York, 1983.
3. G. R. Cooper and C. D. McGillem: *Modern Communications and Spread Spectrum*, McGraw-Hill, New York, 1986.
4. W. R. Bennett and S. O. Rice: "Spectral Density and Autocorrelation Functions Associated with Binary Frequency-Shift Keying," *Bell System Tech. Journal*, vol. 42, Sept. 1963, pp. 2355-2385.
5. W. R. Bennett and J. R. Davey: *Data Transmission*, McGraw-Hill, New York, 1965.
6. J. G. Proakis: *Digital Communications*, McGraw-Hill, New York, 1983.

Chapter 15

Phase-Shift Keying

15.1 Binary Phase-Shift Keying

Binary phase-shift keying (BPSK) is a form of phase modulation in which the instantaneous phase deviation $\phi(t)$ takes on one of two values depending upon the values of the modulating message $m(t)$. Usually the two values for $\phi(t)$ are taken to be symmetric with respect to zero and are represented as $\pm \psi$. The relationship between $m(t)$, $\phi(t)$, and the modulated signal are shown in Fig. 15.1. Careful examination of the figure reveals that three conditions have been implicitly assumed.

1. Each bit interval of length T contains an integral number of carrier cycles.
2. The signal phase for transmitting a mark is shifted by exactly 1 half-cycle (180° or π rad) from the signal phase for transmitting a space.
3. The carrier is phased relative to the bit transitions such that phase shifts always occur when the modulated signal is at zero amplitude.

The net effect of forcing these conditions to be satisfied is that the sinusoidal segment corresponding to each bit begins and ends at the same amplitude value. Thus jump discontinuities in the modulated signal waveform are avoided. When the combination of baud rate, carrier frequency, phase modulation angle, and switching instant are selected to prevent jump discontinuities, the resulting PSK signal is called *coherently switched* PSK. Many books, especially introductory ones, present coherently switched PSK as the only form of PSK

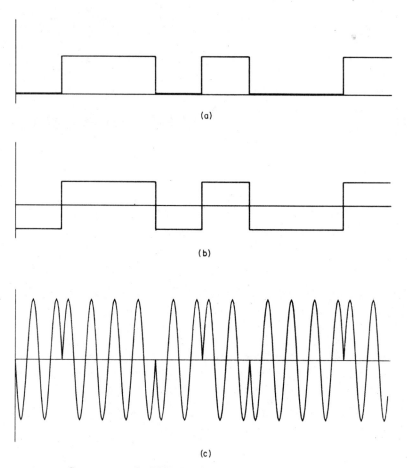

Figure 15.1 Components of a PSK signal. (*a*) Message signal. (*b*) Instantaneous phase deviation. (*c*) Modulated signal.

modulation. BPSK signals which satisfy condition 2 are sometimes referred to as *phase reversal keyed* (PRK) signals.

The three conditions presented above are sufficient to avoid jump discontinuities, but they are not necessary conditions. They can be relaxed somewhat, but since they are related, they must be considered together as a set. Condition 1 can be relaxed to permit an integer number of half-cycles of carrier in each bit interval. There are several wireline modem designs which transmit 1200 bits/sec using a carrier of 1800 Hz, thus providing 1.5 carrier cycles per bit as shown in Fig. 15.2. Conditions 2 and 3 are tightly coupled. The actual condition necessary for avoiding jump discontinuities requires that the bit transitions occur at the instant where the two differently phased

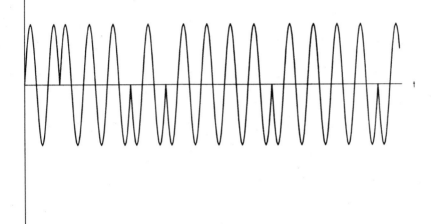

Figure 15.2 PSK signal with 1.5 carrier cycles per bit.

sinusoids cross. When the phase difference is 180° (i.e., condition 2), the sinusoids will cross each other at their zero-crossing instants (hence condition 3). Consider Fig. 15.3, in which the two phases differ by 90°. Here the sinusoids cross at values of $\pm\sqrt{2}/2$ times their peak amplitude.

The signal set for BPSK can be represented by

$$s_1(t) = A \sin(\omega_c t + \psi) \qquad (15.1)$$

$$s_0(t) = A \sin(\omega_c t - \psi) \qquad (15.2)$$

Using trigonometric identities, Eqs. (15.1) and (15.2) can be put into the form:

$$s_1(t) = A \cos \psi \sin \omega_c t + A\sqrt{1 - \cos^2 \psi} \cos \omega_c t \qquad (15.3)$$

$$s_0(t) = A \cos \psi \sin \omega_c t - A\sqrt{1 - \cos^2 \psi} \cos \omega_c t \qquad (15.4)$$

It is often convenient to let $\alpha = \cos \psi$ and rewrite (15.1) through (15.4) as:

$$s_1(t) = A \sin(\omega_c t + \cos^{-1} \alpha)$$
$$s_1(t) = A\alpha \sin \omega_c t + A\sqrt{1 - \alpha^2} \cos \omega_c t \qquad (15.5)$$

$$s_0(t) = A \sin(\omega_c t - \cos^{-1} \alpha)$$
$$= A\alpha \sin \omega_c t - A\sqrt{1 - \alpha^2} \cos \omega_c t \qquad (15.6)$$

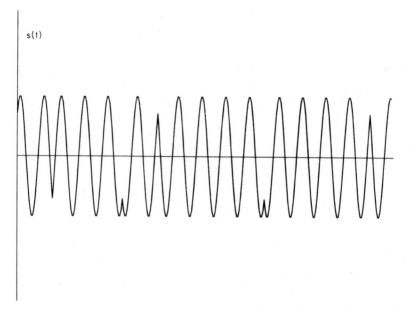

Figure 15.3 PSK signal without jump discontinuities even though bit transitions are not located at zero-crossings.

Relative to these representations, various characteristics of the BPSK signal set are as follows:

Modulation angle $= \psi$ or $\cos^{-1} \alpha$

Carrier component $= A\alpha \sin \omega_c t$

Power in carrier component $= (A\alpha)^2/2$

Modulation component $= \pm A\sqrt{1-\alpha^2} \cos \omega_c t$

Power in modulation component $= A^2(1-\alpha^2)/2$

Correlation coefficient between s_1 and s_0: $\rho = 2\alpha^2 - 1$

15.2 Phase Reversal Keying

Phase reversal keying (PRK) is a special case of PSK in which $\alpha = 0$. The signal set then simplifies to

$$s_1(t) = A \cos \omega_c t \tag{15.7}$$

$$s_0(t) = -A \cos \omega_c t \tag{15.8}$$

This signal set is antipodal [that is, $s_1(t) = -s_0(t)$]. In many texts, PRK is the only form of binary PSK considered. Phase reversal keying can be viewed as amplitude modulation with a bipolar baseband message signal and a modulation index of 100 percent. Therefore, the power spectral density of a PRK signal must equal the power spectral density of an amplitude shift keyed (ASK) signal minus the impulse at the carrier frequency. The psd of a PRK signal is given by

$$G_x(f) = \frac{A^2}{4}[G_m(f-fc) + G_m(f+fc)]$$

where $G_m(f) = \dfrac{\sin^2(\pi f T)}{\pi^2 f^2 T}$

Since the PRK signal set is antipodal, the receiver structures shown in Figs. 10.5 and 10.6 can be used with the resulting probability of error given by Eq. (10.27).

15.3 Geometric Representation of PSK Signals

The modulation components and carrier component for BPSK are in quadrature, so the signals $s_0(t)$ and $s_1(t)$ can be represented as the vector sum of two components as shown in Fig. 15.4a. For phase reversal keying, the signal set has only one basis function (that is, s_0 and s_1 are not linearly independent) so the signal space is one-dimensional as shown in Fig. 15.4b.

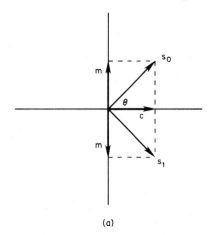

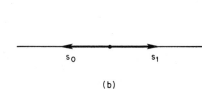

Figure 15.4 Geometric representation of PSK signals. (a) General PSK with modulation components, carrier component, and modulation angle denoted by m, c, and θ, respectively. (b) PRK.

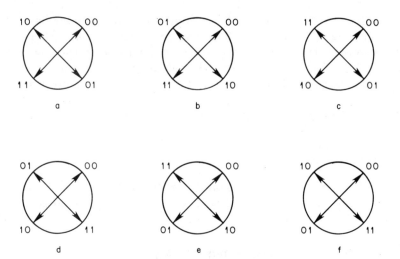

Figure 15.5 Possible dibit-to-phase mappings for QPSK.

15.4 QPSK

QPSK signals can be viewed as two BPSK signals which are orthogonal to each other. Various authors use the abbreviation QPSK to signify "quadraphase shift keying" [Ref. 1], "quadriphase shift keying" [Refs. 2 and 3], "quadrature phase-shift keying" [Ref. 4], or "quarternary phase-shift keying" [Ref. 5]. Excluding rotations, there are six ways to uniquely assign four 2-bit patterns (also called *dibits*) to four phases as shown in Fig. 15.5. When channel noise causes a particular transmitted phase to be incorrectly received, it is more likely to be mistaken for an adjacent phase (that is, $\pm 90°$ from the correct phase), than to be mistaken for the opposite phase (that is, $\pm 180°$ from the correct phase). Depending upon how the dibit values are assigned to phases, an incorrect phase decision will result in either a single bit error or a double bit error. To minimize the overall bit error rate, the dibit values are usually assigned to phase values such that the more likely adjacent phase errors result in only single bit errors. Only the assignments in *a* and *b* of Fig. 15.5 satisfy this requirement. Since the QPSK signal set is biorthogonal, the receiver structure shown in Fig. 10.11 can be used with the resulting probability of error given by Eq. (10.36).

15.5 *M*-ary Phase-Shift Keying

In *M*-ary PSK the binary message sequence is broken into subsequences of k bits each where $k = \log_2 M$. There will be $2^k = M$ unique subsequence patterns, with each different pattern mapped into a different phase. QPSK is a special case of *MPSK* where $M = 4$. Proakis [Ref. 6] distinguishes *MPSK* signals from orthogonal, biorthogonal, and equicorrelated signals by stating that *MPSK* signals exhibit decreasing channel bandwidth requirements for increasing M, while the other three types of M-ary signals exhibit increasing channel bandwidth requirements for increasing M. This distinction may not be readily apparent for $M \leq 4$ since M-ary PSK signals can also be: orthogonal signals when $M = 2$, biorthogonal signals for $M = 4$, and equicorrelated signals for $M = 3$.

A *multiple phase-shift keyed* (*MPSK*) signal set can be represented as:

$$s_m(t) = A \cos(\omega_c t + \theta_m) \quad 0 \leq t \leq T \quad m = 1, 2, \ldots, M \quad (15.9)$$

where $\theta_m = \dfrac{2m\pi}{M}$

Using trigonometric identities, Eq. (15.9) can be put into the form

$$s_m(t) = A \cos \theta_m \cos \omega_c t - A \sin \theta_m \sin \omega_c t \quad (15.10)$$

As shown in Sec. 2.2, the functions $\phi_1(t)$ and $\phi_2(t)$ form an orthonormal set over the interval $[0, T]$ when

$$\phi_1(t) = \sqrt{\dfrac{2}{T}} \sin \omega_c t$$

$$\phi_2(t) = \sqrt{\dfrac{2}{T}} \cos \omega_c t$$

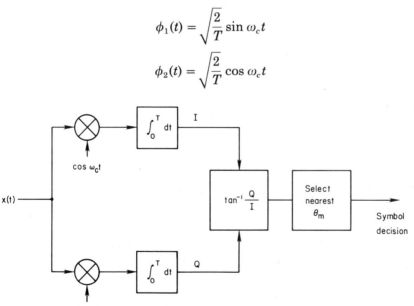

Figure 15.6 Receiver structure for *MPSK*.

Thus Eq. (15.10) can be rewritten in terms of the basis functions ϕ_1 and ϕ_2 as:

$$s_m(t) = \left(A\sqrt{\frac{T}{2}}\cos\theta_m\right)\phi_2(t) - \left(A\sqrt{\frac{T}{2}}\sin\theta_m\right)\phi_1(t)$$
$$= (\sqrt{E}\cos\theta_m)\phi_2(t) - (\sqrt{E}\sin\theta_m)\phi_1(t) \tag{15.11}$$

Equation (15.11) shows that any MPSK signal can be represented as a linear combination of just two basis functions. Therefore, a receiver, such as shown in Fig. 15.6, is equivalent to a receiver which correlates the received signal with each of the m members in the transmitted signal set.

15.6 References

1. R. M. Gagliardi: *Introduction to Communications Engineering*, Wiley, New York, 1978.
2. J. J. Spilker: *Digital Communications by Satellite*, Prentice-Hall, Englewood Cliffs, N.J., 1977.
3. G. R. Cooper and C. D. McGillem: *Modern Communications and Spread Spectrum*, McGraw-Hill, New York, 1986.
4. H. Taub and D. L. Schilling: *Principles of Communication Systems*, 2d ed., McGraw-Hill, New York, 1986.
5. K. S. Shanmugam: *Digital and Analog Communication Systems*, Wiley, New York, 1979.
6. J. G. Proakis: *Digital Communications*, McGraw-Hill, New York, 1983.

Index

Additive gaussian noise, 137–138
Amplitude modulation, 179–183
Amplitude of sine wave plus gaussian noise, distribution of, 124–126
Amplitude of a sine wave with random phase, distribution of, 123–124
Angle modulation, 202–205
Antipodal signal, 152–153
Argand, J. R., 2, 112
Argand diagram, 112
Autocorrelation, 90–91
Autocorrelation function, 60–62
Autocovariance function, 61

Band-limited channel, 138–139
Bandpass noise, 121–123
Bandpass signals, 92–94
Bandpass systems, 115–116
Bayes decision criterion, 147
Bayes rule, 30
Bernoulli numbers, 9
Bernoulli trials, 31–32
Bessel functions:
 evaluation of, 23
 of the first kind, 20
 identities, 20–22
 modified, 22
Binary decision problem, 142–146
Binomial coefficient, 27
Biorthogonal signals, 155–157
Bit error rate, 141
Borel field, 30

Carrier delay (*see* Phase delay)
cdf (*see* Cumulative distribution function)
Certain event, 29
Chapman-Kolmogorov equations, 68
Characteristic function, 37–38

chf (*see* Confluent hypergeometric function)
Chi-square distribution, 54
Circular convergence function, 57
cis, 13
Coherent demodulation, 194–195
Combinations, 27
Complementary error function (erfc), 24
Complex envelope, 93–94
Complex numbers, 11
 conjugates of, 11
 logarithms of, 14
 operations on, 12–13
 polar form, 12
 rectangular form, 11
 roots of, 14
Confluent hypergeometric function (chf), 24, 52
Convolution integral, 102
Correlation, 40
Correlation coefficient, 40
Covariance, 40
Cross-correlation, 91–92
Cumulative distribution function, 33

dB (*see* Decibel)
Decibel (dB), 4–5
Decision criteria:
 Bayes, 147
 ideal observer, 147
 maximum likelihood, 146
 Neyman-Pearson, 147
Delta function, 17–19
Demodulator gain, 187–188
Derivatives, 14–15
 of polynomial ratios, 15
Digitization, 163–170
Dirac, Paul A. M., 2, 17
Dirac delta function, 17–19

Dirichlet, P. G. L., 2, 80
Dirichlet conditions, 80
Discrete Fourier transform, 171–172
Distributions, 18–19
   amplitude of a sine wave plus gaussian noise, 124–126
   amplitude of a sine wave with random phase, 123–124
   chi-squared, 54
   envelope of a sine wave plus bandpass gaussian noise, 126–128
   Erlang, 55
   exponential, 49, 51
   gamma, 23
   gaussian, 45–48
   normal, 45–48
   phase of a sine wave plus bandpass gaussian noise, 128–130
   Poisson, 55
   for postdetection integration of bandpass gaussian noise, 130
   Rayleigh, 27, 49–51
   Rice, 24, 52–54
   squared envelope of bandpass gaussian noise, 128
   uniform, 48–49
Double sideband suppressed carrier modulation, 195–198

e (base of natural logarithms), 3
Elementary events, 29
Energy signals, 76–77
Energy spectral density, 87
Envelope, 92
Envelope delay (see Group delay)
Envelope demodulation, 188–193
Envelope of a sine wave plus bandpass gaussian noise, 126–128
Equicorrelated signals, 157–159
Equivalent noise temperature, 120
erf (see Error function)
erfc (see Complimentary error function)
Erlang distribution, 55
Error function (erf), 24–25, 45
Euler, Leonard, 2, 3
Euler numbers, 9
Euler's constant, 3
Euler's identities, 8
Expected value, 35
Exponential distribution, 49, 51
Exponential functions, 3–4

Field, 30

First passage time, 69
Fourier, J. B. J., 2
Fourier series, 77–84
Fourier transform, 84–89
Frequency modulation (see Angle modulation)
Frequency-shift keying, 219–221
   coherent detection of, 226
   continuous phase, 221–226
   diversity performance of, 240
   M-ary, 231–234
   M-ary, detection of, 234–238
   noncoherent detection of, 228–231
   nonorthogonal, 227–228
   orthogonal, 226–227
   performance in Rayleigh fading, 239
Fundamental theorem of expectation, 35

Gamma distribution, 55
Gamma function, 23
Gauss, J. K. F., 2, 47
Gaussian distribution, 45–48
Gaussian random process, 63
Group delay, 114–115

Heaviside, Oliver, 2, 110
Heaviside expansion, 110

Ideal observer decision criterion, 147
Identities:
   Euler's, 8
   trigonometric, 7–8
Impossible event, 29
Impulse function, 17–19
Impulse response, 101–102
Independent events, 31
Integration, 15–17
Intersymbol interference, 207
Isotropic radiator, 134

Jointly gaussian random variables, 47–48

Kummer's transformations, 24

Laplace, P. S., 2, 103
Laplace transform, 103–107
Linear systems, 98–99
Logarithms, 4–5

Magnitude response, 113
Marcum Q function, 27, 210
Markov, A. A., 2, 64

Markov chains, 64–72
Markov processes, 64
Maximum likelihood decision criterion, 146
Modulation:
  amplitude, 179–183
  angle, 202–205
  double sideband suppressed carrier, 195–198
  frequency, 202–205
  phase, 202–205
  single sideband, 198–200
Modulators:
  Hartley, 200–202
  square-law, 183–185
  switching, 185–187
Moments:
  absolute, 37
  central, 36
  general, 37
  joint, 40

Napier, John, 2, 4
Neyman-Pearson decision criterion, 147
Noise equivalent bandwidth, 118–119
Noise figure, 121
Normal distribution (*see* Gaussian distribution)
Nyquist rate, 166

On-off keying, 207–218
Optimal coherent detection:
  of binary signals, 148–152
  of M-ary signals, 153–154
Optimal noncoherent detection, 159–161
Orthogonal signals, 153–154
Orthonormality of sine and cosine, 9–11

Parseval's theorem, 84
pdf (*see* Probability density functions)
Permutations, 27
Phase delay, 113–114
Phase modulation (*see* Angle modulation)
Phase of a sine wave plus bandpass gaussian noise, 128–130
Phase response, 113
Phase reversal keying, 246–247
Phase-shift keying:
  binary, 243–246
  geometric representation, 247
  M-ary, 248–250
  quadrature, 248

Poisson distribution, 55
Poles of a transfer function, 111–113
Postdetection integration of bandpass gaussian noise, 130
Power signals, 76–77
Power spectral density, 62, 87
Preenvelope, 94
Probability:
  conditional, 30–32
  definition of, 29–30
  joint, 30–31
Probability density functions, 33
  conditional, 39
  for functions of a random variable, 41–42
  for functions of two random variables, 42–43
  joint, 38
  marginal, 38
Propagation:
  line-of-sight, 136–137
  space wave, 136
  surface wave, 135–136

Q function, 25–26
  Marcum's, 27, 210
Quantization, 164–165

Radio frequency bands, 133–134
Random processes, 57–59
  gaussian, 63
  linear-filtering of, 62–63
  uncorrelated, 61
Randon variables, 32
  mean of, 34–35
  moments of, 34–37
  statistically independent, 41
  uncorrelated, 40
Rayleigh, Lord (John W. Strut), 2, 50
Rayleigh distribution, 27, 49–51
Rice distribution, 24, 52–54
Riemann, Georg F. B., 2

Sample functions, 57
Sampling:
  ideal, 165–166
  instantaneous, 167–169
  natural, 169–171
Sampling theorem, 166–167
Schwartz, Laurent, 18
Sigma field, 30
Simplex signals, 158
Single sideband modulation, 198–200

Sinusoids:
  orthonormality of, 9–11
  phase shifting of, 6
Square-law detector, 193–194
Squared envelope of bandpass gaussian noise, 128
State probabilities, 65
Stationarity, 59–60
Statistical independence, 31, 39
Step response, 102–103
Strut, John W. (Lord Rayleigh), 2, 50
Superposition, 100
Superposition integral, 102
Systems:
  additive, 99
  causal, 100–101
  homogeneous, 99
  linear, 98–99
  time-invariant, 100

Thermal noise, 119–120
Transfer functions, 107–109
  poles of, 111–113

Transfer functions (*Cont.*):
  zeroes of, 111–113
Transitional probability, 65
Transorthogonal signals, 157–159
Trial, 30
Trigonometric functions:
  definitions, 6
  product expansions, 9
  series expansions, 8
Trigonometric identities, 7–8

Uniform distribution, 48–49

Variance, 36–37

Wessel-Argand-Gaussian diagram, 112
White noise, 117–118
  gaussian simulation of, 172–177
Wiener-Khintchine theorem, 62

Zeroes of a transfer function, 111–113

ABOUT THE AUTHOR

C. Britton Rorabaugh received his BSEE and MSEE from Drexel University. He is the author of *Signal Processing Design Techniques* and *Data Communications and Local Area Networking,* both published by TAB Professional and Reference Books.